CLIMATE
CHANGED

Creative Director: Saeah Wood
Production & Editorial Manager: Amy Reed
Editorial: Amy Reed, Christa Evans, Holly Lyn Walrath
Design: Ivica Jandrijević
Cover Illustration: Elizabeth Evey
Author Photo: Matt Gough

This book was made with love by humans and does not contain
any AI generated content.

Library of Congress Control Number: 2025904887

Paperback ISBN: 978-1-955671-64-4
E-book ISBN: 978-1-955671-65-1
Audiobook ISBN: 978-1-955671-66-8

OTTERPINE

otterpine.com
Asheville, North Carolina

CLIMATE CHANGED

THE SCIENCE OF SUSTAINABILITY AND HOW EACH OF US CAN DO OUR PART

PABLO RIBEIRO DIAS, PHD

OTTERPINE

CONTENTS

PART 3: THE SOLUTIONS

PREFACE

AT THE TIME of completing the writing of this book, a lot of things had changed in the world, and in my world. Climate disasters have become ever more present. There were heat waves in South Asia and Europe that took the lives of thousands of people, mimicking some (now not-so) science fiction novels. Unprecedented storms in North Africa killed thousands of people and left many more missing. In my home country of Brazil, droughts and fires continue to spread wider and stronger in the north, while elsewhere in the country, there is severe flooding affecting hundreds of thousands of people and literally swallowing whole cities.

I was there when the latest flood happened. I had moved back to Brazil from where I was teaching in Australia to found a start-up. My wife was 40 weeks pregnant when the floods hit and submerged many areas of the capital, where we were living at the time. The neighborhoods to the north, south, and west of us were underwater—the water got so close that we could see it from our apartment. Electricity and water outages soon followed. Several floors of the hospital where my wife was scheduled to give birth were completely submerged, and our obstetrician had to be rescued by an army truck whose wheels made it tall enough to make it through. This rescue took place two days before our baby arrived. We had to find another hospital, taking alternate routes because a good chunk of the roads were impassable. After the baby was born, we sought refuge at my mom's house, which still had water in the water tank. Roads were blocked, bridges fell, and the airport was completely flooded, making it extremely hard to go anywhere. My mother-in-law hopped on many different buses and spent over a week traveling to see her daughter and meet our newborn. While our situation was far from ideal, I saw firsthand far less fortunate women with newborn babies who had lost their homes and were staying in improvised shelters, only surviving thanks to the kindness of strangers who were volunteering.

I have been studying climate change and other environmental problems for a long time and have dedicated my work to making our society a sustainable one. As traumatic as this experience was for my family, it was not surprising to me. But for people around me and in my home state, it was a wake-up call. Climate change was no longer some abstract theory for scientists to talk about and politicians to argue about. It is real and it is now.

This book is about this reality.

INTRODUCTION

I HAD BEEN SEARCHING for a long time for a book that could teach me what I needed to know about the environmental challenges we face in the twenty-first century. I felt constantly inundated with information, yet I didn't have a clear sense of what the threats are exactly. How bad are they? Are they all related to climate change, or are other things going on? While we're at it, how bad is climate change, really? Can we connect all the climate events we see in the news to human activity, or do we need more evidence? What are the ways forward, realistically speaking? What can I actually *do*? Do my actions even matter?

Oh yeah, and I wanted it all explained in simple, plain language that was easy to digest. I didn't want a book by scientists for scientists; rather, I wanted one that was accessible to a mainstream audience, a book that successfully translated complicated topics so everyone could understand them, simplifying as much as possible while staying true to the evidence and scientific consensus. The more I looked, the more I realized this approach was hard to find.

Ultimately, I was unable to find such a book. I read several in the process, but many were too focused on a single country or region—typically North America or Western Europe. Sure, there are many important lessons to be learned from these countries that are ahead in policymaking (and that have way more experience polluting the planet), but physics, chemistry, and biology don't care much about the political borders we have imagined for ourselves. A book addressing the global issue of our current environmental challenges seemed pressing. And I mean *all* environmental challenges, not just climate change, which has become the exclusive focus of many publications and the media in general. While there are many great reads out there, none of them were the accessible, comprehensive, science-based, and solution-focused book about the environment I was looking for. And since I could not find it, I decided to write it.

I am not a climate expert nor am I an environmental scientist—at least not by formal training. I am an engineer who has specialized in waste management and environmental engineering as both a scientist and university professor. Formal expertise is not what I bring to the table. What I offer is quite the opposite: ignorance. I wrote this book out of a huge curiosity, the humility that is necessary for learning, and the passion to make a positive impact. These are traits I believe we all can (and should) possess.

Yet, in the spirit of curiosity and learning, I found myself standing at the intersection of academia and real-world application. I had the opportunity to work as a researcher and as a lecturer in prestigious universities like the University of New South Wales (UNSW; Australia), the University of Sydney (Australia), Macquarie University (Australia), and Universidade Federal do Rio Grande do Sul (UFRGS; Brazil). As a lecturer and researcher at UNSW, my days were spent exploring the intricacies of environmental engineering, renewable energy, and recycling, and teaching students about environmental impacts and how to measure them. My journey, though grounded in engineering, has led me to embrace the broader and ever-evolving challenges of sustainability.

Over the years, I've been fortunate to contribute to a body of work that ranges from developing new recycling technologies to assessing the environmental impact of everyday products like bottled water. Through life cycle assessment—a tool we'll delve into later in this book—I've sought to quantify the unseen impacts of the products we use and the waste we generate. It's a method that's as much about asking the right questions as it is about finding the answers.

But my work has not been confined to lecture halls and research labs. Recently, driven by the same curiosity that led me here, I started SOLARCYCLE, a technology start-up aimed at creating a true circular economy in the realm of renewable energy. It is yet another step that I hope will add to the larger global movement toward sustainability.

While I may not wear the badge of a climate scientist, I am humbled to have been recognized in 2023 by Business Insider as one of the 30 global leaders working toward climate solutions.[1] This recognition isn't about expertise—it's about the ongoing journey to learn, adapt, and, hopefully, inspire others to do the same.

Obviously, my job as a researcher has required me to read a lot about environmental challenges: how to quantify, classify, and tackle them. While I claim ignorance, it's not like I started with a blank page and wrote this book from scratch. But the more I learned about the topic, the more questions I had, and the more I realized how little I knew about it. It is from this teachable place that I attempted to answer the question on so many of our minds: How do we save the planet?

With all this being said, you may be wondering: Is this book scientific? Short answer: Yes, but...

While care was taken to validate all the information presented here with scientific references, it was, as I've mentioned, written in a way that is accessible to a large audience, meaning the writing is more informal and perhaps a little more ambiguous than your average (often excruciatingly dry) science journal article. But even more importantly, this book is biased. I will be the first to admit that I've had past experiences and beliefs that will undoubtedly influence what I write about and how I write about these things: the place where I grew up, the reality I have known, the interactions I have had—all of these have shaped me as an observer and, consequently, shaped the framing of this book. But I'll tell you a secret: Every book (I'll go further—every person) is biased. As a scientist, I try to counter my bias as best as I can by reading things contrary to my (original) belief and checking their validity. This means I am open to arguments, evidence, and facts that may contradict my understanding. It also means I can change my opinion as I learn and study more. So, while I used references and scientific studies to build the narrative and arguments of this book, be assured that as new evidence comes to light, things can certainly change. After all, the basic idea behind science is precisely that we are ignorant beings, so we do the best we can and correct ourselves along the way. This means that I am open to hearing counterarguments and to updating sections of this book if readers provide evidence for it. Please remember that science is the best weapon to fight the ignorance humans have created so far. It is not bulletproof, but it is certainly the most effective.

Throughout this book, I was careful to separate things that were my personal opinion from things that were derived from reliable sources (e.g., scientific papers, consensus reports, credible books, etc.). When something is my personal opinion and based solely on what I make of a given situation, I've made a deliberate effort to indicate that.

All units in this book are presented in the metric system, unless explicitly stated otherwise. To keep the content accessible and easy to follow in both print and audiobook formats, no equations, graphs, or other images have been included. Throughout, many footnotes offer additional context and insights that complement the main text.

This book is split into three main parts. In part I, we'll review how we arrived at both our current environmental crisis and the environmental movement that was born in response. I'll provide a brief overview of how humans have put pressure on our environment to the point of risking ending the human species, and the choices that were made along the way that led us to this potential collective suicide. Were they deliberate? When did we humans start impacting the environment? Have we always been a destructive species, and all that change was population growth, or was there a turning point in history? Moreover, when did we start to notice there was a problem and started to sound the alarm bells? We'll review some of the key recent events and organizations that create the collective awareness of the environmental crisis we face today.

Part II shifts our focus to the core issues of the environmental crisis. Chapters 3 through 5 address global warming, the rapid climate change humanity has observed in recent decades. While global warming is a critical challenge, achieving true sustainability on this planet requires a broader perspective, encompassing other pressing environmental issues such as land use, biodiversity loss, water quality, toxicity, ozone depletion, and more. Some of these concerns intersect with global warming, while others are less directly connected. If we concentrate exclusively on climate change, we risk overlooking other urgent challenges that also demand our attention and action.

In chapter 6, we'll examine these additional environmental threats, though in less depth than our discussion on climate change. This approach is partly to keep the book concise, but more importantly, because climate change has become the defining issue in the environmental struggle. Addressing it effectively requires unprecedented global cooperation—a unity that will also be essential to overcoming other environmental crises. Tackling climate change is a step toward true sustainability, and while it's essential to prioritize this urgent threat, we must also remain vigilant to other critical environmental challenges. And that is precisely what we'll cover in part III of this book: the solutions to climate change and other environmental threats explored in part II.

This book is meant to serve as a guide to what you can do as an individual to maximize your impact—in other words, where to spend your energy, your time, and your money. We'll also discuss the technologies that can assist in fighting the climate crisis and other environmental threats. My goal is not to provide you with a step-by-step guide to environmental activism, but rather to give you the understanding of the key things that need to be accomplished today for us human beings to live in equilibrium—sustainably—on planet Earth.

My promise is that when you finish this book you will be fully equipped with the knowledge you need to separate the environmental wheat from the chaff, so to speak. You'll have a good foundation of knowledge of what climate change is and the science behind it. You'll also understand the many other environmental threats we face today and how they relate to climate change (or not). You'll understand the big picture of the issues we face, as a species and as a planet, and how we can measure the impacts of our actions and those of others. Most importantly, you'll be empowered with the confidence and knowledge to turn your resolve into meaningful action—knowing that together, we have the power to save our planet, literally.

PART I

THE BEGINNING– HOW WE GOT HERE

CHAPTER 1

SCIENTIFIC PROGRESS AND THE AGE OF EXTRACTION

WHAT IS SCIENCE?

YOU PROBABLY KNOW (or think you know) the answer to this question. But before moving on, take a moment to try to define it. If that's too hard, maybe try to come up with a concrete example of something that is science and explain why that is. Go on—I'll wait.

I find it amusing that I went through primary school, middle school, high school, and earned a bachelor's degree without understanding what science really is. I believe I started to understand it while studying for my master's degree, but I only had a full grasp when I was doing my doctorate. The funny thing is that I had science classes during all these schooling phases. I obviously don't know all the school systems worldwide, but I would comfortably bet that in most places, kids are not properly exposed to the meaning of science.

Science, as defined and understood today, is the study of subjects using the scientific method, which is systematic and evidence based. It consists of observation and experimentation, and has to follow a certain set of rules. Put simply, the scientific method is used to answer questions. These questions arise from observing the world around us— what we understand as reality. We then try to question our observations and come up with possible explanations for them, i.e., we reason why something is occurring. This explanation is called a hypothesis. And this questioning and answering can be used for any phenomenon. Really! The only caveat is that the explanation we come up with has to be testable. In other words, there needs to be a way to experiment and prove (or disprove) your hypothesis. Performing the necessary test (this is generally where the bulk of time and effort goes when developing a scientific study) will result in a conclusion either confirming our hypothesis or rejecting it. And the method goes around again, with a different question, a different hypothesis, or a different set of experiments. It's a never-ending process of tweaking and adjusting our understanding of the world.

The scientific method can be used by anyone; it is not exclusive to scientists. However, since it is an activity that requires specialized knowledge, credentials, and resources in order to be formally recognized (and the human species is all about recognition), it's generally undertaken by scientists. But when, for example, a lamp fails to light up your room, you'll probably use the scientific method intuitively. Observation: The lamp did not work. Possible explanations: The power is out, the bulb has burned out, or the wall switch is broken (these are your hypotheses). You test your first hypothesis by turning a light on in another room. If it turns on, the explanation of a power failure is discarded. You then change the bulb and test it. The light turns on. So your experiments led you to conclude that the hypothesis of the burned-out light bulb is correct. The difference with professional scientists is that they would need to repeat such a situation several times, control for things that can interfere with the result, and publish their conclusions in specialized outlets after obtaining their results. By the way, the publishing of results is essential so the scientific community (and the nonscientific community) can see what has been done and why, then base the next set of questions and experiments on those published results.

Another important point is that the results obtained by the scientific method are universal; that is, they should be repeatable by anyone who replicates the proposed method under the same set of circumstances. This is quite important because it decreases the chance of propagating incorrect or unfaithful results for too long. Let's say I came up with a chemical solution capable of making trees grow twice as fast. I then explain my method and my result and publish this data for others to read. If someone in China, Russia, Ethiopia, or Chile tries to copy my method (use the same solution under the same circumstances), they should arrive at the same result. If enough independent groups do arrive at the same conclusions (or similar ones, for that matter), we have strong enough evidence to believe that my magical chemical indeed does as I claimed originally. More often than not, these independent groups will fine-tune my conclusion. Perhaps it only works in specific types of trees, or perhaps trees only grow twice as fast in warm climates, etc.

Another example: Imagine I create a perpetual motion device (one that, by definition, violates the second law of thermodynamics that says energy has quality and spreads out, becoming less useful as we use it). I could explain exactly how I created such a device and the results of the experiments I ran on it, showing that it works. This would definitely attract skepticism. (And it should! Skepticism is an important characteristic of science and critical thinking in general.) Once published, other people around the world could try to replicate my method. If they can't arrive at the same result, they too could publish about their failed attempt and debunk my machine (how sad!). What's more, other people around the world could critique my work and propose that I used the wrong method to test my machine from the start, and show that is why I had skewed results. Like I said, the results have to be universal to be valid. This is beautiful in the sense that it does not matter whether you are a renowned professor in your field or a fresh student learning a subject; the results of your studies are equally valid.

The crucial thing about all of this is that we tend to steer toward the truth (a pragmatic truth, I guess) because science is always questioning itself, and new findings can debunk old theories, which are generally substituted with new theories that better explain the results. And please note this order, because it is quite important: The theory needs to accommodate the results (the evidence), not the other way around. It goes against the scientific method to propose a theory and then look for results that corroborate your theory while dismissing the ones that do not. I call your attention to this because it has been happening too often nowadays.

Perhaps an easy way to picture this is by thinking of a simple web search. While relying purely on Google search results is definitely not science appropriate, we can extract the required insight from this example. If I search for the statement "vaccines are safe," I'll find plenty of pages ("results") that corroborate my theory. If I search for "vaccines are not safe," I will also find plenty of pages backing up that statement. The same goes for searching for "vaccines cause autism" versus "vaccines do not cause autism." The list goes on. A more appropriate search would require one to take a step back and ask, "Are vaccines safe?" without having already made a decision about the answer. Then you could analyze the available data and arrive at the conclusion. (By the way, vaccines are safe and do not cause autism.)

We talked about how science is dynamic, meaning it is always changing. In this never-ending search for the truth, we get it wrong a lot. Sometimes a group of people get it wrong together. But science has been built to correct itself. Remember, the truth is always being updated. In his video entitled "Is Most Published Research Wrong?"[2] from the YouTube channel Veritasium, Derek Muller explores the statistics of false positives, incorrect publications, and the probability of a given piece of research (that used the scientific method) being incorrect. It is a very good example of how science is not perfect, but at the same time it is the best method we know (the least flawed in comparison to everything else humans have developed).

In a similar vein, in his book *Subliminal: How Your Unconscious Mind Rules Your Behavior*, Leonard Mlodinow argues that it is impossible not to be biased.[3] This goes for the scientist, for the job interviewer, for the journalist, for the teacher giving a grade; in short, for everyone. It is up to us to understand this in order to consciously try to minimize these biases. This is a key reason why antagonistic points of view and the scientific method are important: The former prevents us from going in the wrong direction for too long, and the latter allows us to assess this direction in a more objective way. Here, too, the importance of the scientific community stands out because the repetition of studies is a way to reduce our subliminal tendencies to confirm an idea we like, which can cause us to ignore contradictory evidence.

I won't bother making a case for why science is beneficial to society or list everything it has achieved so far (Steven Pinker already did a great job at this in his 600-page book *Enlightenment Now*[4]), but I want you to remember that science is at the core of most of the great developments in the last 500 years, and it creates a single language in which different cultures and societies can speak and exchange information. It brings us closer to understanding our world and agreeing on what is objectively true.

Interestingly, the very science that has catapulted us to the modern life we see around us today is arguably the cause of the environmental threats we face in the world. It gave human beings unprecedented power. And the use of such power brought some unintended consequences.

THE DARK SIDE OF THE SCIENTIFIC REVOLUTION

We have established that modern science is great, but things get complicated when we start talking about climate change and environmental impacts. One of the consequences of the Scientific Revolution was that human technology advanced at a rate that allowed great control over natural resources. We were able to mine ores in larger quantities, pump water long distances, hunt and fish with ease, and farm and harvest with high efficiencies. This was by design. One of the ideas embedded during the Enlightenment and accompanying Scientific Revolution was that humankind was the master of the environment, as opposed to being a part of it. Nearly a century earlier, Francis Bacon, considered one of the forefathers of modern science, advocated for the control and manipulation of nature for the benefit of human beings, which began to be seen as a sign of progress. The idea that nature must be "bound into service," made a "slave," and put "in constraint" by the man of science was entrenched in Bacon's publications.[5,6] It was only much later that the same science realized how bending nature to the benefit of a single species (*homo sapiens sapiens*) can severely damage the environment to the point at which it's not even beneficial to that same species anymore. The intensification of the exploitation of nature initiated in modernity brought with it unique characteristics arising from the social, cultural, and economic conditions of the modern era. We can't be naïve, though, and think that human beings only started exploiting their surroundings in the modern era. There were also pre-modern societies that interacted in predatory ways with nature. However, the potential worldwide environmental cataclysm does not derive from the relationships between pre-modern societies and nature, but from the exploitative relationship of modern society with nature due to rapid and out of control Baconian "progress."

While this historical foray into the development (and misuse) of science is interesting to discuss and be aware of, it does not invalidate or counter the scientific method itself. Noticing the destruction of our environment and potential irreversible consequences is also a feat of science. The same science that evidenced these issues is constantly researching ways to minimize, mitigate, and counter the dire environmental situation we find ourselves in. This book uses scientific findings and arguments to illustrate these challenges and their potential solutions.

ONE WORLD—A PARALLEL BETWEEN THE PANDEMIC AND THE ENVIRONMENT

The COVID pandemic took a heavy toll on society. The personal losses were immeasurable. At the time of writing, global statistics show that two million lives were lost, not to mention the isolation, job losses, rise of mental health issues, loss of schooling for children, distancing from loved ones, and countless other traumas, both shared and personal. I don't think any of us who lived it will need reminding of this in the decades to come.

Can we look on the bright side? Is there one to look to? I am of the belief that the pandemic had the side effect of showing people around the world what globalization really means. It demonstrated in a very real and painful way that we are all living under the same roof, inhabiting the same planet. It stressed that there is no way of fixing global issues solely with local, isolationist solutions. The world is one. Yuval Noah Harari, perhaps one of the sharpest and most influential thinkers of the twenty-first century, published a short piece on the pandemic in which he says that the antidote to this and future pandemics is cooperation. In his words: "We are used to thinking about health in national terms, but providing better healthcare for Iranians and Chinese helps protect Israelis and Americans too from epidemics. This simple truth should be obvious to everyone, but unfortunately it escapes even some of the most important people in the world."[7] From this perspective, nationalism (and its close siblings xenophobia, distrust, prejudice, racism, etc.) is the extreme opposite of what we need to overcome these large-scale challenges. It takes only a single person in a single country to infect all persons in all countries. An epidemic in any country puts at risk the whole human species—can we make a parallel here with the environmental crisis?

The pandemic can also be thought of as a wake-up call. If one nation makes the necessary sacrifices to combat the climate crisis and bears the economic cost, will other nations come together to support it? During the pandemic, we saw several countries resisting lockdown due to fear of economic losses. Perhaps in hindsight we will find that the global toll, in whatever parameter you want to measure, could have been significantly reduced if countries were willing to take drastic quarantine and isolation measures along with international debt-free financial aid. How much would the USA pay, in hindsight, to avoid the disaster caused by the virus? Moreover, will we see a direct correlation between the toll of the pandemic and the lack of cooperation between countries, the lack of belief in specialists, and a lack of faith in the public institutions? The COVID pandemic is a kind of prologue to the climate crisis and many other global challenges to come. Hopefully, we'll be able to clearly see our mistakes and take the cooperation route instead of the isolation route (figuratively and literally) we had to painfully endure during this episode.

CHAPTER 2

ENVIRONMENTAL AWAKENING

L ET'S START WITH some background and briefly describe the events that brought us to the present day, concerning climate change awareness and the green movement. While this summary is by no means intended to cover these milestones in-depth, understanding their origin and impact certainly assists in understanding the big picture.

Many think of environmental problems starting with the Industrial Revolution, but long before then, it was not uncommon for powerful nations to base their entire economies on the idea of extracting as much as they could from a given land and taking all its resources to the main metropole (think of the colonizing exploits of Portugal and Spain in the fifteenth century, for instance). The devastation of the natural Brazilian forest started long before Brazil was dubbed a country. The Atlantic Forest, which hosts one of the largest biodiversities worldwide (and no, I'm not talking about the Amazon), has only about 10%–15% of its vegetation left, and its destruction started over 500 years ago, when Europeans first set foot on Brazilian soil.[8,9] My point is, the destruction of our environment is a topic much older than the Industrial Revolution, and pinpointing an exact starting point would be not only extremely difficult, but also futile to our discussion. However, while the extractive economy of the pre-Industrial era was undoubtedly harmful to the environment, the rate of resource use, waste generation, and emission release grew to a whole new level after the Industrial Revolution, and that is why the conversation about environmental conservation tends to start around the eighteenth century.

The Industrial Revolution started in the middle of the eighteenth century in England, then expanded in an outward radius—and keeps expanding to this day, albeit in a less symmetrical fashion. From the beginning of human societies until the Industrial Revolution, global economic growth averaged 0.01% per year. In other words, the global standards of living were pretty much flat. By the year 2000, average gross domestic product (GDP) per person was 50 times that of what it had been for thousands of years.[10] Given that the GDP is a measure of the goods and services a country produces, we can easily understand why the Industrial Revolution placed so much stress on the environment: It was the beginning of an era during which humans increased their economic activity fiftyfold!

The question at hand is, when did human beings start to notice environmental damage? Or, in other words, when did we start to understand that we can only push the planetary boundaries so far before nature loses its ability to recover and replenish?

It would be ignorant for us to claim that the concern for the environment and the idea of humans living in balance with nature started in the modern era after the Industrial Revolution. Several human societies respected and interacted with the environment in a sustainable manner way before James Watt came up with any steam engine, or the first industries started to show up in Britain. Indeed, there are forms of human societies that have been living in a sustainable manner for centuries and still do to this day (there are plenty of Indigenous communities as examples). They understand that nature requires balance and that extraction can only be pushed so far before the local environment suffers and threatens their way of life. Of course, history also tells us about plenty of other societies that did not act in a sustainable fashion.

So, as we look back in time, it is convenient to point to the Industrial Revolution as the "big bang" of the environmental crisis: It is already considered a historical turning point, and it also marks a point when our ability to cause environmental harm reached an entirely new scale compared to what it had been before. But when did we start to notice that we were indeed causing harm to our *whole* planet? When did this understanding and concern expand from the minds of a few scholars and nascent environmentalists to actually reach a broader slice of society? This shift occurred roughly two centuries after the Industrial Revolution, through the work of the environmental movement's prominent (and not so prominent) founding figures, who stood by the data, organized, and spoke to the world about what they were seeing.

Though environmentalism didn't start becoming mainstream until the late twentieth century, the beginning of what we think of as the environmental movement started in the nineteenth century. This includes the publishing of relevant books on the topic, as well as the creation of important modern conservation groups in the USA and elsewhere, both of which we will discuss further in this chapter. However, the main international treaties would only start in the next century. It goes to show how slow our actions (or should I say reactions?) are when it comes to environmental issues: the Industrial Revolution in the eighteenth century, environmental movements in the nineteenth, international treaties in the twentieth. And now, in the twenty-first century, we are starting to see more significant change in the winds with bigger, bolder, and much-needed action. But we still have long way to go and cannot afford to wait for what the twenty-second century will bring, for we most likely won't be here to see it if we skip our duties in the here and now.

SILENT SPRING

In the mid-1900s, marine biologist and conservationist Rachel Carson made some observations, documented these observations, and eventually wrote a book about them called *Silent Spring,* making history. It was a big deal for two reasons: First, the book correlated human activities and the destruction of ecosystems; second, it popularized the idea that our actions have consequences on the environment and that those consequences may end up being more than we can handle. The publication of *Silent Spring* in 1962 is considered by many to be the beginning of the modern environmental movement. In it, Carson explores the unregulated use of pesticides that were developed in World War II. The book focuses on DDT (dichloro-diphenyl-trichloroethane) in particular, illustrating the ecological imbalances that resulted from its use.[*] Additionally, Carson observed that DDT does not simply disappear from the food chain. It is said that in nature nothing is lost, nothing is created, and everything is transformed. Once crops have absorbed DDT, whatever eats the crops will have DDT in their systems. The more of that crop a creature eats, the more concentrated the pesticide will be in its system. DDT also enters waterways due to rain, contaminating things that will be eaten by bigger things, which will be eaten by even bigger things, which can eventually be eaten by humans. And when humans are harmed by pesticide chemicals, alarm bells ring.

But Carson also proposed that there can't be healthy human beings without having a healthy ecosystem, regardless of whether pesticides reach them directly at the end of the food chain. Toxic chemicals like DDT cause harm to humans indirectly due to the harm to the whole biota that surrounds us (and on which humans are dependent on for

---

[*] One thing Carson observed was DDT's lack of selectivity in what it targeted, showing that it indiscriminately eliminated unintended species that had their own role in the ecosystem. This causes an imbalance that leads to the rapid growth of other species, creating pest situations.

their own lives, by the way). Many have tried to discredit Carson's work. After all, it was in the way of "progress." But careful evaluation by many scientists has confirmed Carson's claims and reiterated her conclusions.

SUSTAINABILITY AND *OUR COMMON FUTURE*

As the idea that human actions could damage the biosphere started to slowly take root, so too did the idea of sustainability. But what exactly is sustainability? How can we measure it? The answer to this is quite complex and will pop up throughout this book. But we can start in the early 1980s.

In 1983, Gro Harlem Brundtland, the former prime minister of Norway, was appointed chairperson of the United Nations World Commission on Environment and Development (WCED), forming what was to become known as the Brundtland Commission. Four years later, in 1987, the Commission released a report titled *Our Common Future*, which had a significant international impact. One of the main features of the document is the definition of sustainable development as "development that meets the needs of the present without compromising the ability of future generations to meet their own needs."[11] This definition would be widely used as a basis for the formulation of documents and regulations concerning public and private directives in years to come—and it is still relevant to this day. The document also recognizes the link between environmental conservation and social aspects, calling for a just and equitable distribution of resources within and among nations (there are mentions of redistribution of wealth, poverty reduction, and gender equality). Decades later, Gro Harlem Brundtland reflected upon the original document and kept the conviction that "poverty is the main cause and effect of environmental degradation in many developing countries."[12]

When speaking of sustainable development, there are two distinct schools of thought: weak sustainability versus strong sustainability. The former presupposes that natural capital and human capital are interchangeable, i.e., it is okay to make use of natural capital as long as it is being converted into human capital. A vivid example would be to say that it is okay to tear down a forest, as long as the wood is being used to build homes for people, thus increasing the human capital. Conversely, strong sustainability does not consider these to be interchangeable, i.e., they are distinctly different forms of capital.

There is a simple rule that states that sustainable development is achieved when such development does not decrease the capacity to provide utility (per capita) over time (this is known as Hartwick's rule). This means if resources are extracted at the same rate as they are replenished, such resources can (theoretically) provide utility virtually forever[13] (even increasing in utility if the resources are "optimally extracted"). Note, however, that this rule does not distinguish natural capital from human capital. So going back to the forest example: If we tear down the forest and use its wood to build houses, and then we tear down these houses to make better houses (supposing technological advancements allowed us to do that), and then we tear down these new houses to build even better houses, and so on, so the capital per person is ever-increasing. From a weak sustainability standpoint, an extraction-rebuilding model like this is an example of a sustainable development. However, what about the species that were unique to the original forest and went extinct once it was torn down? The weak sustainability school of thought says this is still sustainable development because the capital rose overall. It supposes this is a fair trade-off: The extinction was the price to pay for the better houses—and this is okay because the resources are interchangeable. From a strong sustainability perspective, however, the extinction and the houses are not interchangeable resources, because once the species is extinct, future generations cannot utilize this resource (for example, by admiring a species deemed beautiful, or through the harvesting of a specific output only produced by such species*). In this sense, a weak sustainable approach says it's okay to burn fossil fuels (natural capital) as long as the energy obtained from them is put to good use by creating capital (schools, roads, machinery) for humans. In contrast, the strong sustainability approach sees natural capital as nonsubstitutable. It acknowledges that we'll always lose some capacity of the natural capital, whether in its ability to produce goods, absorb pollution (often loosely called "resilience"), or provide environmental amenities (landscapes, shade, clean(er) air, etc.).[14]

Keep in mind, however, that both weak and strong sustainability are still *economic* concepts. Both still assume that natural capital can be monetarily valued.[15] Moreover, both perspectives are based on assumptions that cannot be tested under the scientific method and that make claims about the (unknowable) future.[16] Ultimately, these economic ideas are more complex than portrayed here, but hopefully you get the gist of it, and you can always go off to read more about it if curious.

We have now come full circle, returning to the definition of sustainability proposed in *Our Common Future*. Interestingly, one can argue that this definition allows for the application of both a weak and a strong sustainability framework. Meeting the needs of a given generation can be interpreted in significantly different ways. How much food, water, shelter, goods, etc. does one *need?* Only enough to survive? A bit more? How much more? Where is the line between need and superfluous? Going back to the example where the extinction of a species was justified by the houses that were built, how can we define whether

---

* As you can see in this example, strong sustainability should not be thought of as some kind of idealistic environmental movement. It still treats natural resources as *resources*, meaning their perceived value still lies in their usefulness to humans. Strong sustainability does not mean that no species should ever go extinct or nonrenewable resources should never be used.

these species will be *needed*? It may just so happen that the species in question would be critical for the development of necessary medicine in the future. Let me share a concrete example: Hypertension medicine made from the venom of a snake (the Brazilian jararaca) is a survival *need* for many—and its market value was about $1.7 billion in 2019.[17] How could we have ever known its value (monetary or otherwise) several decades back when the understanding and development of such a drug had just begun? In this sense, it is impossible to justify weak sustainability as the preferred paradigm because one can always argue that an extinct species *can* become a need for future generations. At the same time, it is impossible to prove that a species *will* become a need, also leaving strong sustainability one step behind in having the final say.

Furthermore, at what point will human well-being start to decrease as a consequence of actions taken in the past? On average, human well-being has been steadily increasing over time, but there is plenty of recent evidence pointing toward things going in the opposite direction, with climate refugees serving as a sobering example.[18] It is therefore hard to justify "sustainability" using these limiting economic terms. Or the opposite view can be taken, that it's easy to justify sustainable practices if you just keep redefining what "needs" means. Finally, note that all these concepts are very anthropocentric. That is, when we talk about meeting the needs of current and future generations, we always focus on the needs of human beings. We ignore other species as also having rights for their current and future generations.

THE INTERGOVERNMENTAL PANEL ON CLIMATE CHANGE (IPCC)

The IPCC is a body of the United Nations established in 1988 by the World Meteorological Organization (WMO) and the United Nations Environment Programme (UNEP), and it is also financed by these two organizations. The UN perceived the need for such a panel based on research done earlier in the century that pointed toward the existence of climate change and its possible severe consequences. Both the UNEP and WMO were naturally inclined to take the lead in such a matter, but at the time some key countries opposed this idea. The reasoning behind their opposition was more political than anything: If climate change was as impactful as it seemed, governments would likely want to influence the work directly, instead of relying on "third-party" organizations.* Thus, the idea of an intergovernmental panel was put forward, made up of representatives appointed by member governments.[19] Note that this veered away from a more scientific-oriented organization and toward a political-expert hybrid. Some argue this was a good move, since it allowed for more to be achieved in practice (rather than having only "fruitless research" for research's sake), but the influence governments have in such an organization is always something to bear in mind. Another reason for the push was such a panel would need multidisciplinary experts in addition to climate scientists, such as biologists, physicists, chemists, and a full spectrum of professionals from different areas to assess the causes of climate change, its mechanisms and possible consequences, and mitigation strategies.

The role of the IPCC is priceless, as it aims to assess the changes in climate and provide governments with guidance to create policies to tackle the climate challenge. In 1990, two years after its creation, it released the first assessment report, which, in combination with further documents published later, was used as a foundation for an

---

* The reason being that if you are a country with a big industrial infrastructure, you don't want to be at the mercy of a third-party group that says what you should or should not do (if they say you need to reduce emissions, for example, it will be a burden on your economy). A classic example is that of OPEC (Organization of the Petroleum Exporting Countries) countries that were very resistant to admitting climate change even existed, as it hurt their economies directly (imagine if all of a sudden people stopped buying oil!).

international environmental treaty signed by 154 states at the United Nations Conference on Environment and Development (UNCED), held in Rio de Janeiro in 1992.* This and later assessment reports were used as the base for the Kyoto Protocol (1997), in which industrialized nations agreed to reduce greenhouse gas emissions in accordance with agreed-upon individualized targets.

The production of these assessment reports has multiple stages and involves a vast number of contributors. Field experts and governments are invited at different stages of the process to comment on the scientific, technical, and socioeconomic evaluation of the drafts, which are assessed by hundreds of reviewers. Contrary to what some believe (myself included, at one point), the IPCC does not produce original research or monitor the climate itself. It instead uses available data from the published literature to assess the situation and draw its conclusions. The assessment reports are produced in three parts: (1) the physical science basis for climate change; (2) the impacts and risks of climate change on the ecosystem and biodiversity, with special attention to humans (health issues, societal disruptions, etc.); and (3) proposed mitigation strategies for both the short and long term: how we can do it, how efficient it will be, how much it will cost, etc. The reports not only deal with the technical feasibility of possible mitigation strategies, but also how to enable these strategies from a political-societal perspective.

To date, a total of six IPCC assessment reports have been produced. The AR5 was released in 2014 and was important in arguing in favor of more rigorous policies in the 2015 Paris Agreement, which was a legally binding international agreement on climate change focused on limiting the global average temperature increase. As I was writing this book, the latest IPCC report (AR6) was released in three stages, between 2021 and 2023.

---

* Also known as the Rio Conference, Rio92, or the Earth Summit.

AN INCONVENIENT TRUTH

Of the milestones we'll talk about in this book, Al Gore's 2006 movie *An Inconvenient Truth* was the most influential turning point for me personally. It was the first time I was actually shown the urgency of the climate crisis and the threat it posed to the world. I saw it while at school and it is one of the reasons I later chose to direct my attention toward environmental issues during my years studying engineering.

An Inconvenient Truth was based on a lecture former USA Vice President Al Gore had given many times with the goal of educating people about global warming. The movie discusses several topics that we will go over in this book.* It also sheds light on the role of the media in giving equal attention to science that backs up climate change and to climate change deniers, when there is an overwhelming scientific consensus for the former. Eleven years after its release, a sequel to the movie was released. *An Inconvenient Sequel: Truth to Power* continues the conversation started by the first movie, with updates (the 2015 Paris Agreement, for instance) and with several confirmations of the predictions made in the first movie. *An Inconvenient Truth* is considered a milestone in the climate change fight, not because it portrayed new information, but because it had a great reach and catapulted the climate crisis to capture mainstream attention.

GRETA THUNBERG AND GEN Z

You've probably heard of environmental activist Greta Thunberg, and it's definitely worth talking about why she has gained notoriety in recent years. She is truly independent—not affiliated with any

---

* Including the measurement of carbon dioxide (CO_2) in the atmosphere over time, the CO_2 data from ice cores, statistically significant deviation in temperature patterns, and the decrease in thickness and area of glaciers.

political parties nor backed by big corporations or NGOs. Thunberg
first captured the world's attention by simply sitting by herself outside
the Swedish parliament on a school day in August of 2018. With this
simple act, the then-fifteen-year-old started a series of school strikes
around the world, which spread like wildfire through the media.

But why does Thunberg's activism matter? And how is it different
from so many other memes, videos, and statements that go viral every
single day? For starters, Greta puts her words into action: When
she travels, she always chooses the method with the smallest carbon
footprint, which means trains for short distances and sailing for
long distances. She stuck with her school strike after it was already
a worldwide phenomenon (look up "Fridays For Future"), and she
probably won't stop until the demanded cuts in emissions are met. She
influences people around her to think about the climate challenge and
make changes accordingly; her close family, influenced by her, altered
their ways to have a more sustainable lifestyle. All of these things make
Greta a true leader, not just an internet celebrity.

But those are not the only reasons why I think Greta Thunberg
is an important figure in this story. The reason I believe she deserves
a section of her own is because she is the face of a new generation.
She represents the evolution of the relationship between humans and
nature. It's an evolution that started timidly with the boomers, who
organized themselves, initiated movements, and started bringing the
matter to schools to be taught to the next generation. Gen X and the
millennials grew with the existence of the IPCC, the considerations
of the Rio92 Earth Summit, and the Kyoto Protocol. They further
understood the cause and pushed it forward. But again, they did so
somewhat timidly, and not enough to make the necessary changes
happen. Then Greta came along, equipped with all the information
the new generation needed to see that something is incredibly wrong,
channeling her anger into action. She speaks for an entire generation when she says "You are not trying hard enough" to the people in

power, and that the fault belongs to the older generations, who have failed miserably in their task to give their children a future. This is not, however, an excuse to sit back and watch Gen Z do all the hard work. On the contrary, it is *because* of this generation that older generations need to enact change—now!

While this chapter only skims the surface of the environmental movement, I hope it has given you a helpful overview of where we started, how far we have come, and how much further we still have to go. In the last two centuries, environmentalism has evolved from a niche concern to mainstream prominence, and meaningful progress has been made both in how individuals view their relationship to the environment and how governments accept their responsibility to enact policy to protect it. But I believe history will show that we are currently still only at the beginning, that we have only just started taking the necessary steps toward a sustainable human relationship with the Earth and its resources. Because as you will see in the following chapters, we still have a lot of problems to solve, and perhaps even more importantly, we need the collective will to try.

PART II

THE PROBLEMS

CHAPTER 3

IS CLIMATE CHANGE REAL?

Yes.

I'm glad we cleared that up. Now we can move to the next chapter.

CHAPTER 4

WHAT IS CLIMATE CHANGE?

I DELIBERATELY AVOIDED THE debate about whether climate change is real or not, because no good can come out of such debate. Even acknowledging there's a debate may give the impression that there are two sides with equally valid arguments. The mere repetition of a phrase or idea, even one labeled as false, can mislead people in the long run because *it is easy to confuse familiarity with the truth.*[20,21,22] So much so that politicians have been using this tactic to intentionally create confusion and push their agendas forward—a strategy that has ramped up around the globe over the past ten or so years. The consequences of human actions on planet Earth are visible and measurable. While politicians, economists, journalists, and others may be pushing the idea that the scientific community is still debating whether mankind's activities influence climate change or not, that message is false. There is a scientific consensus on the reality of anthropogenic climate change.[23] This "debate" is similar to the flat Earth movement or the anti-vax movement—they just don't hold up under scrutiny.

Please don't get me wrong—doubting hypotheses, questioning media messaging, and having healthy skepticism are not the issue. Flat Earthers, anti-vaxxers, and climate change deniers have the right to question and should evaluate the available data. They should not, however, create data and only pursue the information needed to prove their point. This is not how science works, and thankfully, science has mechanisms to prevent that, to a certain extent. It is only after many independent researchers come to the same conclusion that a theory is disproven or proven. It's a long road, and the scientific method, while having its mishaps, does tend to balance itself out toward the truth. In a 2016 paper that reviewed several studies on where the consensus stands, the conclusion was that between 90 and 100% of published climate scientists agree that humans are causing the recently observed global warming.[24]

I will, however, make an important remark in this regard. There are indeed issues around climate change that are debatable and that lack a consensus. I will do my best throughout this book to point them out and explain all reasonable sides of the argument and data interpretation. But we have to understand and make a distinction between what is still being debated and what has been overwhelmingly confirmed by evidence.

Remember how we talked about science in the introduction of this book? And how the act of questioning is at the core of the scientific method? Well, this fundamental idea means that scientists are never 100% sure of anything. The very understanding of space and time were redefined by Einstein when he first proposed his general relativity theory. It turns out that this spirit of questioning and constant reevaluation is often mistaken for uncertainty by the public. Yet it is this very same approach that is shaped by finding evidence and by steadily increasing the degree of confidence.

If you are anything like me, you've wondered whether the correct term to use is climate change or global warming. Or maybe you thought global warming is only a part of climate change (or vice versa). It is confusing indeed. What I was surprised to find is that it is confusing *by design*. The original term used was global warming. But in the late 1980s, the USA and Saudi Arabia lobbied to change the term to climate change because it was (at the time) less emotive and less connected to the idea of burning fossil fuels.[25] And so the term "climate change" was used with this political connotation; i.e., the groups (and countries) that adopted it wished to downplay the negative implications for the fossil fuel industry. Eventually, climate activists embraced the term since it is actually more representative of the issue than global warming, as there are indeed far more consequences (and interactions, as we'll see) that are broader and as relevant as the increase in temperature that the term "global warming" suggests. Examples include changes in weather patterns (such as rain, wind, hurricanes), biological activity, atmosphere-ocean exchange, etc. In summary, both terms are somewhat interchangeable, and nowadays, "climate change" is the more accepted and accurate term.

This book, however, aims to go beyond climate change and its direct consequences. We will talk about threats to our biosphere that are indirectly related to climate change, and sometimes even independent of such phenomena. I will make this distinction even clearer as we progress. For now, just keep in mind that when you encounter the term *global warming* in this book, it is making reference specifically to the increase in average global temperature due to human activity, and when you encounter the term *climate change*, it is making reference to the increase in global temperature plus all its other (many) consequences. So climate change describes the whole, and global warming describes only the temperature specific.

ANTHROPOGENIC CLIMATE CHANGE

We know climate change is real because we can measure it. We can also concretely measure its consequences, though some consequences are harder to measure than others. The warming of the globe, for instance, is quite straightforward. Scientists measure the temperature in different locations on the planet on a regular basis. A rise in the measurements indicates that the planet is warming. Interestingly, it turns out we can use proxies to go back in time to estimate the temperature of planet Earth thousands of years ago, before we started to measure it ourselves. The proxies in question can be tree rings (a tree trunk's outward growth from its center); ice cores (long blocks of ice that have been "growing" for thousands of years and freezing certain molecules in the process); corals (whose outer "skeleton" is made mainly of calcium carbonate [$CaCO_3$], which has temperature fingerprints); lake and ocean sediments; fossilized leaves; and many others.

Once we possess such information, we can then graph the measurements with temperature on the vertical axis and time (year) on the horizontal axis. This results in one of the most famous graphs on the subject of climate change: the hockey stick. This graph was originally authored by climate scientist Michael E. Mann and colleagues, so named because the trajectory of global temperature is in the shape of a hockey stick, with a (relatively) flat temperature period (the "shaft" of the stick) to start, followed by a sharp increase (the "blade" of the stick) around the turn of the twentieth century.[26] It was introduced in 1999 and displays the increase in temperature measured in the Northern Hemisphere. It has since been updated with different databases and has remained consistent with a large body of literature that came afterward.[27]

Once we have confirmed the answer to "Is there a warming?", the next question becomes, "Is it natural?" We know temperature does fluctuate due to a myriad of factors, the change in the physical positioning of Earth and Sun being the most significant. Let us start by differentiating the fluctuations. Seasonal variations are indeed shown in the hockey-stick plot. In fact, it only looks like a hockey stick if we take the average temperature of a certain year (or else it looks like a sideways tornado with a very skewed funnel). However, when we take a step back and look at expressive fluctuations—the ones that are observed in geological time scale (that is, millions of years)—we can see that the rate of change is much faster than we'd expect. This fast rate of change is precisely the problem. If the change was occurring in a slow and steady fashion, there wouldn't be reason to be alarmed, as that is actually the default situation. Planet Earth is always changing, and species are evolving continuously. That is one of the main ideas Rachel Carson puts forward in the introduction of her book *Silent Spring*: Nature adapts over time, but because things have been changing so quickly since the Industrial Revolution, there has not been time for it to adapt. We refer to climate change as the rapid and recent change in climate that has been caused by humans (anthropogenic climate change), as opposed to the natural, slow, and ever-occurring climate change that can be traced back millions (excuse me, billions) of years.

So there is scientific consensus about global warming, and the rate of change is worrying, but how do we know it is human made? It turns out that carbon from fossil fuels (which is formed by the transformation of dead biomass and has been the main energy source used by humans for decades) and carbon naturally arising from respiration have different concentrations of isotopes,* the former having more of the lighter isotope (carbon 12) and the latter having more of the heavier isotope (carbon 13). The ratio between the two in the atmosphere (amount of carbon 13 atoms divided by the amount of carbon 12 atoms) has been decreasing over time, meaning that the amount

---

* Isotopes, if you skipped this lesson in chemistry, are atoms that have the same atomic number but a different atomic mass. In other words, they are the same but different, like a kangaroo and a wallaby. Sure, if you ask a biologist (or an Australian for that matter), they'll be able to tell you the differences, but for everyone else it's the same thing (Google them individually and let me know what you learn). Likewise, scientists can tell the difference among isotopes, but each isotope behaves much like its counterpart. Perhaps a better example is two cars that are the same model, make, year, etc. The difference between the two is that one car has four people inside, while the other car has five. The cars are otherwise the same and have the same functions, but one is slightly heavier than the other.

of carbon from fossil fuels has been increasing in relation to natural carbon. And that is how we know for sure it is a human-made change.

GREENHOUSE EFFECT—THE SCIENCE

The greenhouse effect is a concept that is often misunderstood. It will help to start with the basics: The sun is Earth's primary energy source. Almost any type of energy source you can think of is direct or indirect solar energy (a case against this can be made for some sources like geothermal and nuclear, but let's simplify things for the time being). Hydropower is derived from the potential energy that water gains when it evaporates and is deposited on higher ground; the difference in height between the upper and lower reservoir can then be converted into electrical energy, which we use to power our appliances. But where does the energy to evaporate all this water come from? You guessed it, the sun. All the energy contained in fossil fuels (coal, oil, gas) is indirectly related to the sun, too, because fossil fuels are products of organic matter that once grew by extracting energy from the sun (plants) or eating things that extracted energy from the sun (heterotrophic species). We also know the sun indirectly participates in wind energy because it affects the ambient temperature, local pressures, and many other factors that create the windy conditions necessary for kinetic energy to be extracted from the wind and then converted into electrical energy. The most obvious one is solar energy, in which photovoltaic cells take advantage of the incoming photons (packages of energy), striking a semiconductor and creating an electric field to harvest electrical energy. The list could go on, but my point is that the sun is our golden goose.

This amazing star that we call the Sun constantly emits energy in the form of solar radiation [about 340 watts (W) per square meter] to planet Earth. Think of radiation as waves that travel through space.

These waves come in all shapes and forms, have all sorts of bells and whistles, and fall on a wide continuum of wavelengths. But for our purposes, we can simplify all that and just split the waves that come from the sun into two types: long and short.*

The radiation leaving the sun travels through space and eventually finds its way to our planet. When these waves reach our atmosphere, a series of things start to happen. Part of the radiation is reflected by the atmosphere (that is, it's sent back to space), part is absorbed by our atmosphere, and part makes its way to the surface of our planet. On the surface, white bodies (mainly regions covered in ice and deserts with their white sands) reflect the *short*wave radiation, sending it back "upward"; some of these reflected waves will make their way to outer space, while some will be absorbed by our atmosphere (note that this is the second chance the atmosphere has to absorb energy). Clouds are also good at reflecting radiation back to space. Dark bodies (you may read this as "nonwhite" bodies or surfaces, such as forests, oceans, and urban areas) also reflect some radiation, but they mainly absorb it.

But here is the catch: So far, we've only spoken of shortwave radiation. This is the kind that is reflected by the atmosphere and white surfaces. You can picture it bouncing off our atmosphere or planet's surface and flying away from Earth. However, the radiation that doesn't bounce off instead gets absorbed by the surface, and the atmosphere will "energize" the molecules with which it interacts. Think about a mug of water in a microwave. It is initially cold, but after a minute or two, it becomes hot, meaning that it absorbed some of the energy the microwave emitted. The same thing happens with the molecules that interact with the radiation coming from the sun: They absorb this energy and heat up. The problem is that whenever

---

* Wavelength is a spectrum as opposed to a discrete property. Within the "short" category are many wavelengths such as ultraviolet and X-ray. Within the "long" category are wavelengths such as radio and microwave.

something heats up, it ends up having to release that energy somehow. Think back to the example of hot water in a mug—you can place your hand next to the mug and feel the heat. This is because it is releasing the energy it absorbed. This energy release is also radiation, but it is the longwave type. And unlike shortwave radiation, the longwave type is terrible at passing through matter.

Let's reflect on that (no pun intended): The atmosphere is absorbing all this energy coming from the sun. It heats up and it wants to release this energy. But when it does so, it releases longwave radiation, which can't go very far, meaning the atmosphere won't "allow" much of it to escape into space. Thus, more energy will be absorbed in the atmosphere. The gases in the atmosphere that are responsible for absorbing this energy are called greenhouse gases (GHGs). We'll talk more about them in detail later on, but for now, hold your horses.

So, now we have three sources of energy: the sun (direct source of shortwave radiation); Earth's surface (indirect source emitting long waves); *and* the atmosphere, which has absorbed the energy from both previously mentioned sources and is now also emitting its own cut. While all of this may seem complicated, here is what you need to understand: Energy comes to Earth, some energy goes back to space, and some stays "trapped" here on Earth. This whole phenomenon of trapping energy is what we call the greenhouse effect.

This trapped energy is the source of all our problems, right? If we can get rid of it, we are good to go! Find a way to destroy the greenhouse gases and problem solved! Not really. Remember I told you to hold your horses?

Of all the energy coming from the sun, about two-thirds is transformed into heat. If we did not have the greenhouse effect, virtually all incoming energy would leave Earth eventually. Based on some fancy math and the first law of thermodynamics (which, in simple terms, states that energy cannot be created or destroyed, but is otherwise always conserved), we can estimate what the Earth's surface temperature would be in such a case. It turns out it would be about 18°C. Not that bad, right? Not even sweater weather. But there's one detail that might be worth mentioning: It would be 18°C *below zero*! That's right, the average temperature of Earth's surface would be –18°C (–0.4°F) if our pretty atmosphere did not trap some of that incoming energy. So, while you may read about the greenhouse effect as an evil to be fought, in reality it is a natural (as opposed to human-made) phenomenon that pretty much allows life as we know it to exist. It adds those extra thirty-something degrees Celsius that keep Earth's crust's average at the 15°C we have evolved to be accustomed to.

The greenhouse effect is not limited to Earth's atmosphere. It happens in a variety of different instances. A very obvious one is, well, greenhouses (hence the name). Another is when you leave a car parked out in the sun, only to return and scream at how hot it is inside—you've just experienced the greenhouse effect. The exact same principles apply. Energy comes in the form of shortwave radiation (mainly) through the windows and heats up the car inside, but the heat (longwave radiation) cannot get past the glass on its way out. So energy in the form of heat is trapped inside the car. Likewise, if you were ever wise enough to listen to your elders growing up, you may have heard them tell you to keep the windows and blinds shut on hot summer days, as opposed to opening them up. The latter may seem to intuitively make more sense: Open the windows to let the breeze through, right? But your elders were on the right track: If you have the blinds or curtains closed, they will stop a lot of the radiation from getting into the house in the first place, which means you won't have to find a way to let the heat out later.

GREENHOUSE EFFECT—THE PROBLEM

But if the greenhouse effect is a natural phenomenon and is vital to maintain life as we know it, what is the problem? The problem arises from the way we are managing our atmosphere. In very simple terms, we're releasing more greenhouse gases into the atmosphere, which accentuates the greenhouse effect, which in turn causes the average global temperature to rise (hence global warming).

You may say, "But my friend told me the amount of human emissions is insignificant when compared to the amount that is naturally released constantly." Your friend is an idiot. They're not necessarily wrong, but if the conversation stops there, your friend is an idiot. The science is much more complicated than that.

Imagine you just filled a bathtub with water. The water fills almost the full volume of the bathtub. Now open the drain and also run the tap, so the amount of water going in perfectly matches the amount of water going out. Brilliant, you just achieved bathtub equilibrium! You are pretty much at bathtub-Gaia.* Alright, as long as the situation is maintained, the bathtub will never overflow. As a matter of fact, the water level will stay the same. However, imagine that a very light rain starts to fall into your bathtub (okay, maybe your bathtub just turned into an outdoor jacuzzi...just play along). The amount of water added to the tub from the rain is very small (so small a certain friend of yours would call it insignificant). But after a while, you notice the water level rise a bit. Just a bit, nothing to worry about. The rain keeps falling, though. And the level keeps rising. Thankfully, the tub had a little space at the top, so it could still accommodate some excess water. But the rain continues to fall.

What will happen? You got it: Eventually, the water in the tub will overflow. You may think this is no big deal. Just get a mop and dry the floor, right? If you've been paying attention, however, you'll know that overflow is actually a metaphor for environmental disaster and

---

* If this joke is lost on you, the Gaia hypothesis, developed in the 1970s and named after the Greek Earth goddess, postulates that our planet is a naturally self-regulating superorganism.

pretty much the destruction of Earth as we know it. To make matters worse, what started as a gentle rainfall has been gradually increasing despite international agreements to keep the storm under control. In summary, yes, your friend is correct, but they are also an idiot for disregarding that extra amount of "water" building up in our Earth-tub metaphor. And if you still think that "all we need to do is to come up with a mop," just wait until chapter 9 when we talk about solar radiation management.

If we continue with our Earth-tub metaphor, one can ask if it wouldn't be wiser to "enlarge the drain" instead of waiting to clean up the overflow with a (very dangerous) mop. Ah, yes, the drain; in this case, it means removing the greenhouse gases that are currently in our way. Are there ways to do that? Yes, there are. Some of the modern human-made examples will be explored in chapter 9.

But there is this ancient designer that was really onto something; this designer goes by the name of nature, and it came up with several drains (in science talk, these are known as "sinks"[*]). One is called a tree, and we don't have to develop it because it already exists! As architect William McDonough puts it, "Imagine this design assignment: Design something that makes oxygen, sequesters carbon, fixes nitrogen, distills water, makes complex sugars and foods, creates microclimates, changes colors with the seasons, and self-replicates. And then why don't we knock that down and write on it?"[28] Another form of carbon sink is called an ocean. In fact, Earth has been incredibly resilient to the extra carbon we throw at it because of these natural sinks. They give us a buffer between CO_2 emissions and temperature rise. In rough numbers, only about 60% of CO_2 from fossil fuels ends up in the atmosphere; 20% is absorbed by land (mainly in the form of forests) and 20% by oceans.

---

[*] A carbon sink is defined as a natural or artificial mechanism that absorbs and stores atmospheric carbon.

The idea that land and oceans absorb much of the CO_2 we release into the atmosphere is an important concept that will appear throughout this book. These carbon sinks help maintain equilibrium by absorbing excess CO_2. The high concentration of CO_2 in the atmosphere creates a gradient—essentially a difference in concentration—which drives carbon sinks to absorb CO_2 and restore balance. Nature inherently moves toward equilibrium. Later, when we explore methods to remove carbon from the atmosphere, you'll see the reverse effect: These carbon sinks may begin releasing stored carbon back into the atmosphere to balance the new gradient created when atmospheric CO_2 levels drop.

HELLO ATMOSPHERE, IT'S A PLEASURE TO MEET YOU

Sometimes it's hard to make sense of what we are told about climate change because of the scale at which things happen. Changes seem to come slowly, and pollutants seem so small in relative scale (e.g., one part in a billion) and so large in absolute scale (e.g., billions of tons!).[*] Indeed, while carbon dioxide and other GHGs can cause major problems, they represent only a tiny fraction of the air surrounding us at this very instant.

Our atmosphere (the troposphere, to be more precise[†]) is made mostly of nitrogen gas (78%). This is followed by oxygen gas (21%), which is much needed for our breathing and also known as photosynthesis by-product (plant waste). So just these two make up 99% of the atmosphere! The next is the noble gas argon (0.9%), followed by all trace gases (carbon dioxide, hydrogen, krypton, neon, helium, etc.). Trace

---

[*] I'm going to use "tons" a lot in this book. Whenever you see "tons," it will always mean "metric tons," even if "metric" is omitted. A metric ton is 1,000 kg.

[†] The troposphere is the lowest region of the atmosphere, extending from the earth's surface to a height of about 6–10 km (3.7–6.2 miles).

gases are less than 0.1% of the atmospheric composition, and within this small fraction, carbon dioxide is king: the most abundant of the least abundant gases in the atmosphere (0.04% of the total atmosphere, to be a bit more precise). So all this greenhouse gas fuss is about a ridiculously small increase in a ridiculously small number? Short answer: Yes.

MORE ABOUT GREENHOUSE GASES

Carbon dioxide and its equivalents

My guess is that carbon dioxide is the first gas that comes to mind when you think of greenhouse gases. While CO_2 is a natural result of things like animal respiration and volcanoes, it's also the most common toxic by-product of human activities, such as the burning of fossil fuels, so it's generally seen as the big villain in the global warming fight. This perception is not at all wrong, but it is also a bit unfair. You see, one of the reasons why carbon dioxide is always associated with global warming is because we tend to convert *all* greenhouse gases into carbon dioxide equivalents (CO_{2eq}) for the sake of simplification.[*] Methane, for instance, is a GHG that is about 25–28 times more powerful than carbon dioxide in regard to its greenhouse effect during a 100-year time horizon.[†] Thus, we say 1 kg (kilogram) of methane is equivalent to 25–28 kg of carbon dioxide, or 25–28 CO_{2eq} for short. You may see this value change greatly depending on how you compare the gases (e.g., methane is 34, 84, or 120 times more potent than CO_2). This

---

[*] To do that, we normalize (convert) other GHGs so that CO_{2eq} is the amount of warming that one ton of the gas would create relative to one ton of CO_2 over a certain defined amount of time.

[†] A 100-year time horizon means we are considering its impact for 100 years. This is arbitrary and other time horizons can (and are) chosen for different analyses.

is because methane does not hang around the atmosphere as long as carbon dioxide does (it "lives less"), meaning it has a very strong short-term effect, but the effect is weakened when dispersed over a long time horizon. Furthermore, methane and carbon dioxide have different molecular weights, so comparing 1 kg of each will render a different result than comparing molecule for molecule. Yet another difference: If we take into account the feedback loops[*] involved in the emission of gases, the CO_{2eq} values will change (we'll discuss these feedback loops in coming chapters). So, the bottom line is that depending on how you measure it, the CO_{2eq} value will vary. For the sake of this book, we won't worry too much about the precise numbers. What we do need to understand is that the different GHGs have different global warming potentials and different lifetimes in the atmosphere.[29,30]

In addition to carbon dioxide (CO_2) and methane (CH_4), the primary GHGs are ozone (O_3), nitrous oxide (N_2O), and water (that's right, good old H_2O). The Kyoto Protocol also includes sulfur hexafluoride (SF_6), hydrofluorocarbons (HFCs), and perfluorocarbons (PFCs).[31] As I write this passage, total global greenhouse gas emissions are about 51 billion tons of carbon dioxide equivalents per year. The conclusion here is that carbon dioxide, while famous, is not the only villain. It is your "vanilla" GHG. And we use it as a ruler to measure the other gases, both with respect to how potent they are (their radiative forcing[†]) as well as to how long they will last in the atmosphere (their time horizon).

---

* A feedback loop, in simple terms, is a cycle where the results of something you do affect what happens next, often creating a continuous loop of action and response, e.g., when you start getting good at a sport, getting good then motivates you to play it more, which makes you even better, and that makes you even more motivated, and so on.

† Simplified, radiative forcing is a measurement of the ability a gas has to absorb energy from incoming radiation. The same radiation ("sunlight") can either be super-absorbed by a molecule, be somewhat absorbed, or be completely ignored.

Methane

We briefly touched on methane (CH_4) when discussing how CO_2 equivalency is calculated, but we haven't talked about where it comes from. Methane is the second most important GHG after CO_2, accounting for a significant portion of the registered warming. Methane emissions have risen sharply since 2007 (with a particular acceleration from 2014), and there is still great uncertainty around the emission contribution by natural sources versus anthropogenic sources (e.g., wetlands versus fossil fuels and agriculture). This is partially due to the current lack of long-term monitoring data.[32] While the distribution of emissions might not be known with precision, it is known that anthropogenic sources include the coal, oil, and gas industries (e.g., leaks from natural gas wells, coal extraction and production), waste management (e.g., landfills, biomass burning, sewage disposal), and agriculture (e.g., rice paddies and digestion by ruminants such as cows).[33,34,35] A high-level overview of the main sources of methane includes agriculture, waste, wetlands, and fossil fuel production and use.[36] Fires were also believed to be a major source of methane, but it turns out their contribution is around one order of magnitude smaller (in other words, approximately one-tenth) than that of wetlands,[37] which are a major source of natural emissions but can also be affected by human activity.[38]

Ozone

Ozone in the stratosphere is great! But this is some 50 kilometers above the Earth's surface. Closer to us, ozone is bad news. Later on, we'll talk in more detail about ozone's role in keeping ultraviolet (UV) radiation away from us, and also about its role in air pollution and harmful consequences to human health, but for now, let's focus on ozone's role as a gas that contributes to the greenhouse effect. I anticipate this will cause some confusion, as ozone will pop up in different sections of this book, and you might think, *Wait, haven't we talked about this already?* So here is a quick guide to help you make sense of what is to come: Ozone plays three different roles depending on how far away from the surface it is located. If it is close to the ground (in the lower troposphere), it acts as a strong oxidant that can damage goods and be harmful to plants and animals (including humans!) above certain thresholds. If it is a bit farther away, in the upper troposphere (sort of "halfway" between the ground and the stratosphere), it acts as a greenhouse gas, contributing to global warming. If it is far away from the surface, in the stratosphere, it acts like Earth's sunscreen, blocking UV radiation.* To help make your mental image even clearer: The air pollutant role is played in the lower troposphere, where birds fly and we live; the GHG role is played in the upper troposphere, where the helium balloons that have escaped birthday parties end up; the sunscreen role is played in the stratosphere, where long-haul planes fly.

Ozone in the troposphere (also known as ground-level ozone) has a lifetime of a few weeks and originates either in the stratosphere (and "migrates" down in what is called stratosphere–troposphere exchange), or it is produced in the troposphere by precursor gases (nitrogen oxides [NO_X], carbon monoxide [CO], and non-methane volatile organic compounds [VOCs]) in the presence of sunlight.[39,40] Apart from the VOCs, the other precursors are predominantly human made and closely related to industrial activity. Note a difference here with respect to the other GHGs: The anthropogenic emissions are not

---

* None of these roles are exclusive. The ozone molecule behaves the same in all three roles it plays. What changes is not the ozone itself but the consequences of what ozone is doing. So, it will play a role as an oxidizing agent in all levels of the atmosphere, but in the upper parts of the atmosphere, that role does not have any harmful consequence. In the lower troposphere, however, it causes human health issues. This simple three-role explanation is useful to more easily understand ozone interactions and consequences. We will go over each of these throughout the book.

ozone directly, but rather ozone's precursor gases. The origin of these precursor gases is similar to the sources of other GHGs mentioned previously: fuel combustion, emissions from vehicles, power plants, industrial boilers, refineries, chemical plants, etc.[41,42] We'll talk more about ground-level ozone formation and its origins later in the book when exploring air pollution.

The amount of tropospheric ozone is likely to be increasing in the Northern Hemisphere. However, confidence among the scientific community regarding ozone levels is medium to low due to limited measurements and because levels vary greatly with season, altitude, and location.[43] Ozone is often considered the third most important GHG because its radiative forcing (i.e., how much it transforms energy into heat) is third only to CO_2 and CH_4.[44]

Nitrous oxide

Nitrous oxide (N_2O), widely known as laughing gas, is another important GHG understood to be 300 times more potent than CO_2[45] (remember that the exact value depends on the time horizon, unit of comparison, etc.) and it stays in the atmosphere for long periods, 116 years on average. As is the case with carbon dioxide and methane, nitrous oxide is released naturally into the atmosphere as part of the nitrogen cycle. The major natural sources of nitrous oxide are soils (e.g., forests and savannas) and oceans. However, there are also anthropogenic sources that correspond to some 40% of the total current emissions.[46] These include agriculture (through use of fertilizers), biomass burning, power plants, wastewater treatment plants, combustion engines, and nitric acid production.[47] Current estimates are that the concentration of atmospheric N_2O is increasing at a rate of 2% per decade.[48] If we stopped here, nitrous oxide would seem quite similar to other GHGs already discussed. But here is the catch: Even if we were to magically stop the use of fossil fuels today, the anthropogenic emission of nitrous oxide would still go on. This is because we would still require great amounts of fertilizers (synthetic or natural manure) to keep up with food production. And mind you, these emissions are a sign that the fertilizer is working! I am not talking about the emissions associated with the making of fertilizer, but rather the emissions related to the microbial metabolism of nitrogen compounds, nitrification and denitrification, which produce NO and N_2O, respectively.[49] We don't need to understand these processes in detail; all we need to know is that they are related to the transformation of nitrogen fertilizers into nitrous oxide, and that they are one of the main reasons behind the third agricultural revolution (also known as the green revolution), which was the period in the twentieth century that saw huge rises in food yield per area of land. Reducing N_2O emissions in agriculture can be achieved through better management of fertilizers, new fertilizers that lower N_2O production, and alternating crops to include those that do not require fertilizers.*[50]

CO_2 is still the most important of the GHGs, but its contribution in comparison to the other GHGs has been decreasing. These other GHGs have much lower concentrations than carbon dioxide, but are able to absorb more energy from incoming radiation. Just to give you an idea, CO_2 is measured in ppm (parts per million), while CH_4 and NO_2 are measured in ppb (parts per billion, a unit of measurement a thousand times smaller!).

Water

Perhaps the GHG that is most often forgotten by the general public is water. Indeed, water vapor is the largest contributor to Earth's

---

* For instance, plants in the *leguminosea* family (including legumes like peas, beans, and peanuts) have nodules capable of fixating nitrogen from the air, which can eliminate the need for nitrogen in the soil, and so the use of fertilizers.

greenhouse effect.[51] It's a special case, though, because water does not stay in the atmosphere for long; it condenses or solidifies often and falls back to the Earth (as rain and snow). As a matter of fact, water remains in the atmosphere for only a matter of days, while carbon dioxide stays decades or centuries (though a 100-year time horizon is often used as a standard). This short-lived atmospheric life means that water vapor does not build up as CO_2 does. We would not have a greenhouse problem if water was the only substance responsible for converting shortwave radiation into heat; since atmospheric vapor does not build up, there would not be an increase in temperature. This is why water is a special case and often not mentioned when the GHG discussion shows up.

However, water is not insignificant because global warming can allow it to concentrate more in the atmosphere, not necessarily because it lingers longer, but because a warmer climate changes the saturation threshold of the atmosphere. Here we see one of the feedback loops described earlier: The greater the temperature, the more water can be retained in the atmosphere (meaning more water molecules can "dissolve" in the same volume of air), which will then increase the temperature even more, which allows even more water to be retained in the atmosphere, and so on and so forth. This concept is quite interesting, because it means that the greenhouse effect of water is a *consequence* rather than a *cause* of climate change. More water means a hotter world, but it does not increase the amount of carbon in the atmosphere. So, the feedback loop stops if increases in carbon stop. If you remove carbon from the picture, water cannot close the feedback loop on its own. (You may want to read this paragraph again to fully understand this concept, as it is counterintuitive for many.)

Another characteristic of water that makes it different than the other GHGs is that it forms clouds, and to some extent, the clouds have a cooling rather than a warming effect. Some clouds have high reflective properties, which means a portion of the incoming radiation from the sun is reflected and never makes it to the part of the atmosphere, where it would be trapped by bouncing off GHG molecules. All the aforementioned reasons make water unique among the greenhouse gases. This is why it is often disregarded when ranking the GHGs in importance.

NATURAL COOLING: COUNTERFORCES TO GLOBAL WARMING

According to the climate models used to project global warming and what society should do about it, carbon emissions should have peaked in 2010 and decreased thereafter to keep (limit) the temperature increase below the famous 2°C above pre-industrial levels. This did not happen. Instead, we are currently heading toward an even worse scenario (potentially up to a four-degree increase). And because we failed to hit peak emission a decade ago, we now face a procrastination penalty, which means that the longer we take to reduce emissions to be below the required threshold by better mitigating the current situation (i.e., the longer we *procrastinate*), the harder it will be to stabilize GHG concentrations in the atmosphere. Stabilization means bringing emissions to a peak and then lowering those emissions subsequently. The procrastination penalty means that the longer we take to reach the peak, the faster we'll need to decrease emissions in the following years.

The current target, set by the IPCC in the 2015 Paris Agreement, is to keep atmospheric CO_2 concentration below 450 ppm. In 2021, we were at roughly 414 ppm. However, this number only takes carbon dioxide into account, not all the other GHGs we talked about earlier. As a matter of fact, if we did take them into consideration, our CO_{2eq} concentration would be well above 450 ppm, in the vicinity of 500 ppm![52] Does that mean it's too late to keep the warming below dangerous levels? Not really. Just like all those gases that contribute to the heating of the planet, there are some other substances that contribute to cooling it. And it turns out that we humans have also been releasing a fair amount of these substances, too. The main example here is sulfur dioxide (SO_2), which comes from aerosols, the burning of coal, and combustion of diesel fuel or heavy oils (oh, the irony!). It can also come naturally from the eruption of volcanoes, for instance. The effect of sulfur dioxide in the atmosphere is increased reflection of some of the incoming radiation from the sun, essentially decreasing the energy absorbed by the planet. However, scientific understanding of the full effects of this molecule, its quantity, and its lifespan are still unknown. This will be important to remember when we talk about geoengineering (and solar radiation management) in chapter 9. Moreover, while SO_2 may seem like the hero here, it is a major air pollutant that can be harmful to humans and animals, and it can also contribute to the formation of acid rain. The planting of trees—in addition to the benefits from carbon capture and storage due through photosynthesis—also has a cooling effect due to the increase in the amount of energy that Earth's surface bounces back upward. So reforestation initiatives also have the cooling-down effect that counters the greenhouse effect.

Coincidentally, the substances causing the cooling effects pretty much offset one to one the additional warming effects from the GHGs, other than carbon dioxide. Therefore, taking only the concentration of carbon dioxide in the atmosphere by itself is a pretty good indicator of where we stand in anthropogenic global warming, and so we are back to the 414 ppm we started this section with.

When talking about efforts to stabilize emissions, it's important to understand that there are many different ways to frame the global warming problem, which can be confusing. Let us try to detangle the possibilities and have a clear understanding of each concept. The first is the concentration of CO_2 in the atmosphere, the ~414 ppm I mentioned earlier. It is a concentration just like the amount of sugar in a cup of coffee or the alcohol content in a can of beer. Next up is the concentration of CO_2 plus other GHGs (the CO_{2eq} we defined earlier) in the atmosphere, which is the ~500 ppm.

Next up is the infamous 51 billion metric tons that starts and conducts the whole storyline in Bill Gates's bestselling book *How to Avoid a Climate Disaster: The Solutions We Have and the Breakthroughs We Need*. This is the total amount of CO_{2eq} we emit globally on an annual basis. But this is not a fixed number. When we reach the goal of net zero emissions by 2050 (and we better reach net zero by 2050!), this means we will have brought the 51 billion number down to zero, and we will have effectively stopped emitting GHGs into the atmosphere. But note that the CO_{2eq} concentration in the atmosphere may very well stay at the same 500 ppm it is currently even if we hit net zero, as it's not going anywhere for a long time (or not without some kind of huge intervention). This is why you'll always hear about us trying to *limit* the temperature increase, and never[*] about us trying to *decrease* it.

It's also possible you'll find the 51 billion metric tons of CO_{2eq} expressed in different units. Examples include 51 Gt (gigatons) or 51 Pg (petagrams). It's all the same thing. There is also the 37 billion tons of CO_2 we emit annually—but this only includes CO_2, not CO_{2eq}. So just as we jump from 414 ppm to 500 ppm when going from CO_2 to CO_{2eq}, we jump from 38 billion tons per year to 51 billion tons per year when going from CO_2 to CO_{2eq}. It can be very confusing for bystanders to have these numbers thrown at them.

---

[*] Never say never. In the last part of the book, we'll look into ways to actually lower the temperature. But they are quite risky and not realistic at this point in time. Moreover, the pandemic brought a slight decrease in emissions, from 37 to 34 billion tons per year. It shot right up afterward, though.

WHERE DO GREENHOUSE GASES COME FROM?

There are several ways one can group greenhouse gas emissions, much like we broke down the different types of CO_{2eq} in previous sections of this chapter. By doing so, the distribution changes. For instance, agriculture and land are used mainly to produce food, representing 20% of the total emissions. But if we decide to group the whole food system (including the emissions of the packaging industry, transportation, retail, etc.), the figure jumps to 25%.[53] Almost three-quarters of global emissions are related to energy, but grouping as such makes it hard for us to have a clear mental picture of which sectors are meant by "energy." I bet some of you thought of electricity, while others may have thought about fuel for transportation. In *How to Avoid a Climate Disaster*, Bill Gates separates GHG emissions as such: 31% arise from making things, 27% from generating electricity, 19% from growing plants and raising animals, 16% from transportation, and 7% from keeping places warm or cool. Look closely and you'll see an important distinction made by this grouping: while making things requires electricity (to run the factories, power some processes, etc.), this is accounted for in the electricity category. Likewise, an automobile requires fuel, and the making of fuels has its own carbon footprint, but this is accounted for in the making things category. Therefore, transportation only includes the contribution from the activity itself, as is the case with growing plants and animals. These distinctions are often overlooked and get tangled up if not properly defined because the categories will always have overlaps. So let us quickly go over each one.

Making things

Making things has to do primarily with the making of the materials used in bulk by our current society, namely cement, steel, plastics (polymers), aluminum, paper, fertilizers, glass, etc. The making of these materials requires electricity to operate the factories and heat to allow some process to happen, both of which release GHGs as a consequence. In addition, the making of materials in itself may release GHG as a by-product of the very chemical reaction necessary to make the material in the first place. Steelmaking, for instance, revolves around taking iron oxide (the state in which iron is found when mining its ore) and reducing it to metallic iron. This is generally done using a carbon source, so the carbon and oxygen (oxygen from the iron *oxide*) combine. And can you guess the name of the lovely couple formed from carbon and oxygen? You guessed it, carbon dioxide (CO_2). Similarly, producing cement requires calcium, which we primarily obtain from limestone ($CaCO_3$). If you look closely at the chemical composition of limestone, you'll see the problem. Separating the calcium oxide (CaO) from the limestone ($CaCO_3$) will produce CO_2 as a by-product on a one-to-one ratio.

Electricity

Electricity is the second major source of GHG. Today, about two-thirds of all generated electricity worldwide comes from fossil fuels (mainly coal, followed by natural gas, then a small fraction of oil). The remaining third is split among hydropower, nuclear, wind, solar, and other renewables (in order of decreasing representation).[54] Note the distinction between the energy mix versus the electricity mix. The former includes electricity plus heating and transport. So, the gasoline you use in your car is considered part of the energy mix, but not the electricity mix.

Most people learned that using fossil fuels to make electricity contributes to the emission of GHG. But why? Well, fossil fuels are pretty much hydrocarbons (molecules mostly made up of carbon and hydrogen). Gas and coal power plants generate electricity by using hydrocarbons. These power plants work by using these fossil fuels as the source of energy to heat up fluids (generally water or air) that will, in turn, move a turbine. Think of a kettle sitting on top of a burning log. The heat released from the log increases the temperature of the water inside the kettle, which goes from liquid to vapor (steam). The steam then accumulates and raises the pressure of the kettle to the point that it pops the lid open. That is a steam power plant in a nutshell. The coal takes the role of the log, and the popping of the lid is the turbine moving (which will then power a generator). So we take the energy from the coal (or other fossil fuels) and *convert* it into electrical energy. However, since coal is made up mostly of carbon, the burning of coal releases heat (which we want), and also water and CO_2 as by-products (which we don't want[*])—and that is the source of the GHG emission when making electricity.

Have you ever heard that natural gas is better for the environment than coal? This is correct but misleading. Indeed, natural gas has less carbon in its molecular structure than coal, which in turn means it releases less CO_2 for the same amount of energy compared to coal. But it still releases CO_2! It makes little sense to invest heavily in natural gas power plants today if we are aiming to get to net-zero carbon emissions in a couple of decades.

Agriculture

The agriculture category is the next major carbon source. In a nutshell, we need agriculture to feed the human population, but we also use it as a way to obtain several raw materials used in our products (think of wood, textiles, or paper, for instance). Raising animals, growing crops, and harvesting trees are all activities with a significant carbon footprint. These greenhouse gases arise from deforestation, agriculture, and animal

---

[*] Remember the combustion chemical equation: hydrocarbon + oxygen = carbon dioxide and water (e.g., $C_3H_8 + 5 O_2 \rightarrow 3 CO_2 + 4 H_2O$). There are other emissions related to the burning of fossil fuel, which we will discuss in detail in later chapters.

production. Modern agriculture relies heavily on fertilizers to enhance the growth of crops. The main elements in fertilizers are nitrogen and phosphorus. But plants can generally not absorb 100% of the nitrogen that is introduced, and part of the excess finds its way into the atmosphere as nitrous oxide. You'll remember from our earlier discussion that nitrous oxide is a potent GHG. This is one of the reasons that agriculture, through the use of fertilizers, contributes to global warming. One way to get around the nitrogen emissions from the making of fertilizer is to produce clean hydrogen (hydrogen obtained using renewable energy electricity), which can be combined with nitrogen from the air to produce the nitrates.[55] This is technically feasible, but we are still a long way from such a scenario being widespread.

Note here a distinction based on the way we grouped the emission sources: If we're splitting them into making things and growing things, the global warming contribution of fertilizers will also be split between the *making* of fertilizers versus the *use* of fertilizers, which falls in the latter category. Agriculture also has additional impacts on the environment that are not directly related to global warming (land use, water use, pollution, toxicity), which we will address in later chapters.

Animal production is a world of its own. First, feeding livestock requires the growing of crops that we have just discussed, so animal production will always include the impact of growing crops that livestock consumes. And boy do they eat! A 2013 study estimated that if we take the mass of all the crops grown globally, two-thirds are grown for direct human consumption, while the remaining third is mainly to feed livestock (a portion is also used in industrial processes and for the production of biofuels).* If you calculate to consider calories instead of mass, 55% of crops are used to feed humans directly, while 36% are used to feed livestock.[56] In addition to the impact of the crops required

---

* Keep in mind that a portion of crops grown are not edible to humans but are to animals.

to feed livestock, whatever remains after their digestion also has a large footprint. That is to say, animal manure (also known as poop) releases GHGs, nitrous oxide being the main one. Among the various forms of meat animals, the suborder of the ruminants (cattle, sheep, buffalo, goats, etc.) have the greatest impact because their digestive systems produce methane as a by-product. So as these animals are chewing away at their food, they are burping lots of potent GHG. And after their digestion, they also fart away a bit of potent GHG.* These additional impacts are why cutting down on meat is one individual action you can take to reduce your carbon footprint—but we'll discuss just how much of an impact it makes, and the controversies around the subject, in the third part of this book.

Transportation

Next on the list is transportation, which is quite straightforward. We need to input energy (generally in the form of fuel) to allow an engine to output work. An engine converts this fuel energy into kinetic energy (motion). The vast majority (over 90%) of fuels used in transportation are fossil fuels, gasoline being the most widely used, followed by diesel and others (natural gas, jet fuel, kerosene, etc.).[57] When we use these fuels as a source of energy, the by-product of the energy conversion is GHG. Fossil fuels, as mentioned earlier, are basically molecules of hydrocarbonates. Given that we are putting carbon into the tanks of our vehicles, and given that mass is conserved,† we'll be putting carbon

---

* In fact, in a groundbreaking piece of policy, Denmark will implement an emissions tax on livestock farmers starting in 2030, essentially taxing them for cow farts.

† We can apply the continuity principle to our vehicles, which states that the amount of fluid passing through a system must remain constant. It is based on the law of the conservation of mass, which is the idea that mass cannot be created or destroyed.

in and we'll get carbon out. In other words, we are not destroying the carbon that goes into the tank; it is chemically recombining inside the engine, so it eventually leaves, mainly in the form of CO_2 (big surprise, huh?). But in addition to the CO_2, other GHGs such as methane (CH_4) and nitrous oxide (N_2O) are also released through the exhaust of combustion engines. And it does not stop there, because additional substances are also released as a by-product of combustion engines, many that have environmental consequences other than global warming.

Going back to CO_2 emissions, how do we measure the amount of carbon dioxide being released? It is quite simple, actually: The amount of carbon in has to be equal to the amount of carbon out. In more complex terms: We use chemical stoichiometry* and some basic mathematical operations. The IPCC has laid the groundwork by providing default emissions values for different fuels. The amount of CO_2 emitted from transport vehicles, for instance, is calculated taking the sum of all the fuel consumed multiplied by an emission factor, which is simply the carbon content multiplied by 44/12.[58,59]

What about flying? Isn't flying the great climate change villain? Well, not really. It's bad alright, but its overall global contribution is about 1.9%–2.4% of the global emissions [60,61] (so even among the different transportation modes, flying is only responsible for about 11%–15%). Keep in mind this portion is associated with the emissions coming from the release of carbon due to the fuel, and is not inclusive of the making of the vehicles or fuels, which falls under the first category listed (i.e., making things also includes making fuels). So why is flying so demonized in the view of the public? Because the per capita contribution is big! In spite of the overall contribution being (comparatively) small, only a tiny fraction of the world population is responsible for

---

* Stoichiometry measures reactants and products before, during, and following chemical reactions.

it (obviously it is those who are better off and can afford to fly—and within that group, those who can afford to fly multiple times and long distances). We'll get to this discussion later in the book when we talk about what each of us can do. For now, the idea is to give you a sense of where the contributions come from and call your attention to the fact that flying is definitely important, but not as important as you might have originally thought.

The final tally when we sum up all these GHG sources is that infamous 51 billion tons of carbon dioxide equivalents per year, which are the main cause of anthropogenic climate change. So what exactly does that mean, and why is it such a big deal? We'll dive into the consequences in the next chapter.

CHAPTER 5

THE CONSEQUENCES OF CLIMATE CHANGE

W<small>E CURRENTLY HAVE</small> a clear understanding of many of the consequences of climate change. At the same time, there is a lot of uncertainty and speculation around this subject. Therefore, it's important to distinguish the facts from the myths and the "mights." I'll try to do that in this chapter. There are some things that can lead to catastrophic results when it comes to climate change, and I am sure you hear this a lot. But what does *catastrophic* mean in the first place?

A catastrophic event is one that occurs abruptly and with severe consequences (disastrous is a good synonym). But how does that make sense when we know for a fact that the global rise in temperature, for example, is happening slowly and gradually? In material science, we use the term "catastrophic failure" when bridges fall, when ships are torn in half (literally), or when a crack propagates at unmeasurably fast speeds. It is what happens after the breaking point, when a given structure simply snaps. I encourage you to search for catastrophic structure failures to see for yourself—it is scary! But we know that in most cases, before the catastrophic failure occurs, it has a beginning, followed by a steady rise, and then it reaches the final catastrophic event. In other words, a small crack forms, slowly propagates and increases in increments that are barely noticeable if we're not paying attention, and eventually results in an unsteady and catastrophic ending.

Once the science of this crack growth was better understood, engineering got around these failures by routinely inspecting for cracks and assessing at which stage of growth each crack formed. This allowed us to not only foresee when a possible catastrophic failure would happen, but to also take measures to correct it if need be (change a certain component in an aircraft, for instance). The parallel I'm trying to draw with climate change is we know there is a "crack" and that it is growing. We know this crack can lead to catastrophic events. But we don't know exactly where the breaking point will occur. There are so many interactions and unknowns in climate models that it is uncertain how and when things will happen. But climatologists do have predictions, and even if they are not spot on, we ought to listen, for they're the best we have.

One thing we know for certain is that feedback loops can greatly accelerate the growth of the problem, i.e., make climate change occur much faster than current rates. If you recall from the previous chapter, feedback loops happen when the consequence of an event triggers another event that, in turn, triggers the former event again, in a loop. An example is when you sleep and after sleeping you feel hungry: You eat, but after eating you feel sleepy, so you could go back to sleep and let this feedback loop run forever. If you think this is ridiculous, I encourage you to closely observe the behavior of teenagers and you'll see this eat-sleep loop in action. Jokes aside, I have a better example: If you have ever been to a music concert or used a microphone, you may have experienced that very loud noise people refer to as *feedback*. The idea is the same: The microphone is capturing sound and amplifying it to a speaker monitor. If the microphone then captures the sound coming from that speaker, it will trigger a feedback loop in which the sound being amplified (the one that leaves the speaker monitor) is also fed back to the microphone. If you ever experienced this, you know how quickly it escalates.

Going back to the more serious issue of climate change feedback loops, scientists have identified some of them and worry that as the feedback loops accelerate, things will get catastrophic much quicker. One of these feedback loops was already mentioned: water vapor. You may recall that it is a powerful greenhouse gas and that higher temperatures allow air to hold more water vapor—more vapor, more heat trapped, more water in the atmosphere. We'll see other examples in this chapter. The consequences of some events can also stack up, such that one single event can have multiple environmental impacts. Take the case of deforestation and its effect on global warming. Plants are responsible for sucking carbon out of the atmosphere and trapping it in their structures. When they decay, they release CO_2, effectively becoming a source of carbon (plants are made mostly of carbon and hydrogen). They also stop absorbing carbon from the atmosphere once they are dead (obviously). So the more forests are cut down, the more carbon is emitted into the atmosphere (through plant decay), and the less carbon is removed from the atmosphere (through photosynthesis).* This chapter will explore these and other examples of the consequences of global warming.

SEA LEVEL RISE

Perhaps one of the most well-known consequences of global warming is the rise in sea level. Not long ago, it was extremely hard to actually measure small changes in sea level. You can imagine the difficulty, given the changes in tides and constant movement of water. This measuring is generally done with a tide gauge, which is not that different from a

---

* The effect of deforestation proves even more powerful if we remind ourselves that planting forests actually has a stacked effect in favor of global cooling: the carbon capture from photosynthesis plus the increase in albedo, which we'll talk about more in chapter 8.

ruler, but a bit more automatized. Nowadays, however, we have ingeniously come up with a way to measure sea level using satellite data. This means that we have had precise measurements for about three decades. The data shows a clear rise in sea level of about 3.4 mm per year (imagine trying to manually measure the 3.4 mm change with a glorified ruler!). But why does this occur?

The first reason that comes to mind for a lot of people is the melting of the ice caps. And they would be right, but I would bet most of them are wrong about the link between the rise in sea level and the melting of the ice. Most people think the problem is ice melting and pouring into the ocean as water. More water, a higher sea level, right? Not really. You see, if the ice is on land and melting, then yes, I'd agree with that statement. It would melt on land, become liquid, and flow into the sea. However, if the ice is already below sea level, then its melting would not have much effect on the change in volume. A classic way to visualize this is grabbing a cup of your favorite beverage and placing a couple of ice cubes in there. Note the height of the volume occupied by the liquid-plus-solid mix (beverage plus ice). Wait a few minutes for the ice to melt and track the change in height (if any). The ice was already occupying volume in the solid state, so when it melts, the liquid "replaces" that solid volume. As a matter of fact, in the case of water, ice occupies a slightly higher volume than liquid water so you might even see that original height decrease ever slightly if you have a precise way to measure it. On the other hand, if we think of the tip of an iceberg that is above sea level, that fellow will indeed contribute to the rise of sea level, but ultimately, the contribution either way (slightly upward or downward) is only marginal (on average, about 90% of an iceberg is underwater).

The real issue causing sea level rise is actually water expansion! The first time I heard this fact, I was quite surprised (as you might be) and at the same time a bit bummed that I did not think of it on my own. I was well aware of the basic scientific principle that things expand when you apply heat to them, an awareness you may also share, even if more intuitively than consciously. We are surrounded by examples of things expanding when we apply heat to them. Those of us of a certain age may remember mercury thermometers, where temperature was measured by the expanding of mercury as it heated up in that little glass tube. Everyone's familiar with what happens to a pot of water with a lid on it when the water starts to boil: The lid starts to rattle as it's pushed up due to the thermal expansion of the heating water, which creates pressure buildup inside the pot. You might have also noticed the joints commonly used in bridges (these are what make your car go *thump-thump-thump* as you drive across). Bridge joints are special gaps or flexible connections placed between sections of a bridge to handle changes in temperature. When the weather gets hot and the bridge expands, these joints allow the bridge sections to move a little without causing damage or buckling. In cold weather, when the bridge contracts, the joints accommodate the shrinkage, ensuring the bridge remains safe and functional (if you haven't noticed these joints before, make sure a bridge has them before trusting it). Even potholes in roads can be a consequence of the thermal expansion as road materials crack under the increased pressure. Fact is, things occupy more volume when they increase in temperature. Therefore, if global temperatures are rising, that means the temperature of the oceans is rising, which means their volume increases and sea level rises.

As a matter of fact, the rising of sea level is something we can't stop in the near future. There is something called committed climate change, which, as the name suggests, are changes to which we are already committed and cannot avoid. The rise in sea level is one of them. This is because the process of heat transfer from the surface of the ocean to the depths is slower than the transfer of heat on land, and the 1°C temperature rise already observed in land regions has, therefore, not yet reached all the layers of the ocean. This is mainly because of the difference in depth between ocean and land (the ocean is way deeper). So even if today we magically cut off carbon emissions completely, we would still see the increase in sea level for years to come. This delayed consequence is already underway; it is a committed rise in sea level, a committed climate change.

Another source of sea level rise is groundwater. We have become very good at reaching groundwater. But as we use it for agriculture, industry, energy generation, etc., the water that was once underground now flows directly into water streams that often end up in the ocean and contribute to the rise in sea level. Note how this is not directly related to global warming. If we were to magically stop global warming today but maintained the same use of groundwater, we'd still see this water making its way to the oceans.

MELTING ICE SHEETS

Going back to the ice issue, remember when we talked about how the melting of ice does not significantly contribute to the sea level rising? The caveat here is that this is *currently* true. There is, in fact, quite enough ice above sea level to make a difference *eventually*. Specifically, the melting of the West Antarctic ice sheet and parts of the Greenland ice sheet may contribute significantly to the rise in sea level. The uncertainty is in the time scale; in other words, when this melting will take place. Originally, it was thought it would require centuries, but the latest data shows it could take as little as a single century.

Furthermore, you'll recall that we talked about how white bodies reflect a lot of the incoming energy from the sun (in technical terms, this means they have a high albedo, or ratio of reflected energy to arriving energy). And yes, ice is mostly white. So that means portions of Earth's surface that are covered with ice play a role in maintaining the Earth's temperature since they stop part of the sun's incoming energy from being absorbed in the first place. So what happens when ice melts?

Well, the more ice melts, the less ice we have covering the surface, and the less surface we have covered with ice (i.e., less high-albedo surface), the less radiation bounces back.* The less radiation bounces back, the more radiation is absorbed (remember, energy is always conserved); the more radiation is absorbed, the greater the temperature. I wish it would stop there, but notice the positive feedback loop we just got ourselves into. The chain stopped with "the greater the temperature," and what is the consequence of greater temperature? More ice melting...so the loop starts again and spirals down.

Is there scientific consensus around this issue? Yes, but the whole answer is a bit more complex: While the feedback loop I described is a certainty, we don't know how catastrophic it'll be. Don't get me wrong—we know it'll be bad, but how fast it will happen is uncertain. It is possible that the feedback effect will be so strong, we'll see catastrophic results very soon, like within a few years, and it is also possible that loop will feed itself slowly on a downward spiral whose effects will take longer to be noticed, perhaps more on the order of a few generations. It's also worth noting that the melting of ice is not the only thing that can cause this loop. Since it's the surface of the ice that reflects radiation, if we were to cover the ice (or any other high-albedo surface) with something else, the reflection of radiation would be compromised. Smog (arising from industrial activity, transportation, etc.) can land on high-albedo surfaces and compromise the rate of energy reflection. So keeping the whys and hows in mind is useful to avoid losing track of the possible causes of this loop.

Alright, so we know the melting of ice is a big deal due to the loss of its reflective properties and the addition of water to the ocean from land-based ice sheets, but there are even more issues we need to address.

---

* FYI, it's not that all the energy that "bounces back" goes into space. This reflection off the ice is another chance for the atmosphere to absorb the energy.

The same way the moon is always around because it is attracted by Earth, water is attracted by large masses of ice. Wrap your head around this for a moment. Newton's gravitational law says that mass attracts mass; the bigger the mass, the bigger the attraction. Currently, we have massive portions of ice covering the Earth's crust. This ice is so massive that its gravitational force is significant. So you can picture these ice bodies like magnets that pull the ocean water toward them. What happens as ice melts? Well, the magnet becomes weaker and pulls less water. What's so bad about that? It means that the rise in sea level we covered earlier won't be evenly distributed. Regions close to these melting ice masses will be flooded much sooner than regions that are far enough away to not feel the difference. This phenomenon may not be so intuitive at first, but just remember this: Regions close to great ice masses (close to Greenland, for instance) are more susceptible to flooding because they'll receive a disproportionate amount of water. This water is already in the ocean as I write this, but it's "tilted" toward the land due to the mass of ice on land pulling on the water like a magnet. Just like the moon pulls on the oceans and controls the tides, the land also pulls on the water and dictates its level.

PERMAFROST AND MORE

At this point you may be thinking, *Are you done with the ice already?* I wish I could say yes, but unfortunately, there are some additional concerns. There is trapped methane (CH_4) in Earth's icy regions that could be released as they melt. As a potent GHG, this release of methane would create yet another positive feedback effect. But there is more. Picture all the organic matter that sits under the regions that are covered with snow all year round (called permafrost, short for permanent frost): plant material, animal remains, peat, etc. While the snow is present, microorganisms can't reach such matter to decompose it. But what happens if the ice melts? Theoretically it will be decomposed, which will contribute to the greenhouse effect by emitting additional carbon into the atmosphere. The decay of organic matter is a natural part of the carbon cycle. Carbon goes into the atmosphere during decomposition and out of the atmosphere as organic life grows (think of trees and all the work they do). But the matter covered by the permafrost does not participate in this cycle, since it is literally permanently frozen. Will this newly released carbon be recaptured by new vegetation that previously could not grow because of the permafrost? Will the carbon balance be positive or negative or neutral? We don't know. There are too many interactions happening all at once for our models to predict what will occur with precision.

If I made you uneasy with the permafrost problem, hold tight as I am about to tell you about yet another ice melting problem, which involves even more uncertainty. We are a generation who has lived and survived the COVID-19 global pandemic. Do you remember how it all started? Simply put: It was a coronavirus that was completely foreign to all of our immune systems and one day made its way to a human being. We know for certain there are different kinds of viruses and other pathogens trapped in a frozen state in remote regions of the globe.[62] What are the chances one or more gets unleashed when the ice melts? How disastrous could the consequences be? We simply don't know. These are valid scenarios, but there is little certainty around how the chain effect will occur, or even if it will occur.

OCEAN ACIDIFICATION

Carbon dioxide and water react, and the product of such reaction is carbonic acid ($CO_2 + H_2O \rightleftharpoons H_2CO_3$). In an equilibrium situation, you'll have water, carbon dioxide, and carbonic acid all coexisting (i.e., not all the carbon dioxide or water is consumed to produce carbonic acid, so there is still some of each left over). So, no need to panic just yet, because there has always been carbonic acid in the oceans as planet Earth has naturally transformed over time. But when we supply additional carbon dioxide to the atmosphere, there is an increase in the reagents,[*] which results in an increase in the production of carbonic acid. So the greater the atmospheric CO_2, the more CO_2 the ocean will uptake, the more CO_2 is available for the reaction, and the greater the acidity of the oceans.[†] You have probably had a taste of such acid—literally. Carbon dioxide is used to make soda drinks (cola, tonic water, etc.); the fizziness comes from the carbon dioxide bubbles, and the acid taste comes from the carbonic acid that is produced due to the CO_2 plus water reaction.[‡] You can actually find heaps of homemade videos on the internet of people teaching you how to unclog the sink or the toilet using soda drinks. The reason why this works (sometimes) has to do precisely with the acidity of these beverages, though I don't particularly recommend any of these methods.

The 2015 IPCC report states that the ocean has absorbed about 30% of the human-emitted carbon dioxide. Moreover, by tracing the ocean acidity back to pre-industrial levels (which is equivalent to measuring the ocean's pH from the eighteenth century onward) and comparing with today's levels, it was found that the acidity increased about 26% in that period. We are essentially turning our oceans into a fizzy drink, which may appeal to some, but does not appeal to the ocean's inhabitants. Theoretically, ocean acidification should strongly affect marine ecosystems,[63] but here we have another issue without

---

[*] Reagents are the starting materials or ingredients in a chemical reaction. They are the substances you mix together to create something new. Think of them like the ingredients in a recipe.

[†] By the way, the carbonic acid remains dissolved in the water.

[‡] Phosphoric acid (H_3PO_4) is also often added to soda drinks and contributes to the acidic taste.

consensus: How bad will it be? It could be catastrophic if the effects of such acidification are lethal for several species. But it might turn out not to be lethal for many.[64,65] At this stage it looks like acidification makes already existing problems even worse. In coral reefs, for instance, the crisis due to rising temperatures, pollution, overfishing, and introduction of pests is accentuated by the acidification of oceans.[66] One thing we do know with high confidence is that ocean acidification will increase for centuries if the emission of carbon dioxide continues.[67]

EXTREME WEATHER

Extreme weather events are naturally occurring phenomena such as tornados, blizzards, hail, hurricanes, and flooding. Climate change has the potential to exacerbate some of these phenomena, as well as initiate the occurrence of such phenomena in regions where they previously did not happen. The former can be dangerous because the structures in place (buildings, warehouses, roads, bridges, etc.) are not ready to resist the intensified impacts, the latter because there might not even be any infrastructure in place to start with! Humans are particularly susceptible to these events since our bodies require a specific set of conditions to thrive, and countless other species are also at risk.

The changes in weather patterns are not easily relatable to changes in the climate—sometimes the change observed in extreme weather patterns can actually go in the opposite direction of the mean change.[68] An example can illustrate this better: While the average (mean) temperature of the world is increasing, the temperatures in some specific places on Earth are decreasing (which in turn can increase the occurrence of extreme blizzard events, for instance). The exact reason for this requires complex climate models that consider the interaction of several variables, which are out of the scope of this book (and of my knowledge, if we're being honest).

Given the uncertainty on the matter, let us stick to the facts in which we have greater confidence and that were categorized as such by the IPCC. In their words: It is "*virtually certain* that increases in the frequency and magnitude of warm daily temperature extremes and decreases in cold extremes will occur through the 21st century at the global scale" (emphasis added).[69] First, *virtually certain* is the highest level of confidence you will ever hear out of the mouth of a serious scientist. It is when there is evidence that allows you to be sure, but you don't claim it is 100% certain and leave that 0.01% probability of it being incorrect because you know how probabilities work. Next, note that an increase in the *magnitude* is expected, which means that the hottest days will be even hotter, and the coldest days will be even colder.* Then, note that the *frequency* will increase as well! This means that these abnormally hot and cold days should become more common (happen more frequently) than they are nowadays. The IPCC report goes on to say that it is very likely that the duration of heat waves will increase over most land areas, i.e., the episodes of heat waves will be longer in duration too (will last longer). They also report it is likely that the frequency of heavy rain events has increased over the twentieth century and will increase further in the twenty-first century in many areas of the globe.† The "good news" doesn't stop there. There is also high confidence that we'll see more glacial floods and avalanches! Some regions will observe severe droughts while others will experience increased flooding. Droughts happen because warmer soils lose more water. So even regions with increased precipitation can end up seeing droughts. As for floods, the

---

* I'm using the word "days," but it's hotter/colder both days and nights!

† Again, this information is easy to misunderstand when we look only at the general ("average") trend. The report is saying that there should be more episodes of heavy rain on average. This does not mean that all given regions will definitely experience more rain episodes, because the report is reporting on the average of all regions around the globe. A region can actually experience quite the contrary: much less rain, as in severe drought.

warmer the air, the more water it can hold, and hence, more rain (recall the positive feedback loop discussed earlier in this chapter). So all these events that at first seem to cancel each other out can actually be exacerbated with the changes in climate. No fun.

The consequences of these events for us humans, in addition to the obvious destruction, also include indirect effects such as water contamination, the spread of infectious disease, and trauma and mental health issues. And remember, both extreme cold and extreme heat are detrimental to our health. Risk of cardiovascular and respiratory mortality increases in extreme heat events and droughts, but the risk of mortality also increases in extreme cold events.[70,71] Extreme heat is causing countless unreported deaths in poor countries and communities around the world.

Another notorious example of extreme weather events is wildfires, which can be exacerbated by the effects of climate change, both in terms of frequency and magnitude. Here in Australia, in 2019–2020, we experienced what was later dubbed the Australian Black Summer. The name refers to the clouds of black smoke that could be spotted in the sky during those months and whose origin were bushfires of unprecedented magnitude. The event killed hundreds of people due to smoke exposure, with thousands more being admitted to hospitals for heart and lung issues, and billions of animals died. Australia is used to dealing with bushfires, but in the summer of 2019–2020, instead of 2%–3% of the forests burning, 21% did![72] As with many things climate related (and as already stated multiple times in this book), the factors and interactions at play are very complex, and thus clearly linking the extraordinary bushfires with climate change is a very hard task. But an attribution study did just that, at least to the extent possible.*

---

* An attribution study is one that precisely aims to link two or more things. In other words, it is one that assesses whether a phenomenon can be *attributed* to another.

The study found that known climate change consequences (such as long-term warming trends) increased the chances of the fires by at least 30%.[73] But no one should be surprised, because evidence of such phenomena had been piling up: Climate models assessing the bushfires in Australia in the summer of 2018–2019 (one year before the Black Summer) already pointed to an "increased likelihood of dry spring conditions with enhanced anthropogenic greenhouse gases."[74]

In short, climate change has promoted these extreme weather events, and will continue to do so: more heavy precipitation, more floods, more blizzards, more heat waves, more wildfires, more hurricanes, etc., increasing the risk of morbidity, mortality, and diseases. Malaria, a tropical disease distributed by mosquitos, is a classic example of such disease spread. The increase in temperature in higher latitudes expands the climate where disease-carrying mosquitos thrive and thus increases the possibility of people in those regions contracting these diseases. Thus, and in fancy words, a poleward spread of tropical diseases can also be expected with increased climate change.

WATER, FOOD, AND FARMING

After laying out these first sets of consequences of climate change, it should be obvious that there will also be changes to our water reservoirs, food availability, and farming processes. This is a big deal as we currently have about eight billion people in the world to feed and expect to have even more in the next decades (we will run an analysis on population growth in later chapters). How exactly climate change affects water, food, and farming is complicated. Here again, there is a lot of uncertainty, but the latest climate models point to some (troublesome) probable outcomes. And, again, there are many interactions here as the already complex interaction network of climate models meets the complex interaction of food production (agriculture, farming, animal science, etc.).

One of the expected consequences is that we will have both more and less water. Wait, what? Yes, that's correct. As we've discussed previously, some regions are expected to suffer from severe droughts while others from flooding. Needless to say, both situations are terrible for most farming activities. This is also relevant to the maintenance of livestock, as they depend on both water availability and crops. In the words of the IPCC: "There is medium confidence that droughts will intensify in the 21st century in some seasons and areas, due to reduced precipitation and/or increased evapotranspiration."[*][75]

The harvesting of seafood is not safe either. Considering exclusively the climate change–related consequences,[†] there is already plenty to worry about. The disruption of ocean currents, the rise in water temperature, and ocean acidification, all of which are related to climate change, are detrimental to sea life and have already been shown to impact seafood harvesting.[76,77]

Mainly, the models predict a change in the microclimates of different regions. Put simply, regions that today can grow some crops won't be able to anymore, while other regions will become fertile land. Specifically, a growth in crop yield in subtropical regions is expected, but a decrease is expected in tropical regions. Regarding rainfall changes, we'll likely observe more rain near the equator and the poles and less in the mid-latitudes.

The reasons behind the change in farming yield and crop growth are many, going beyond temperature and water availability. For instance,

---

* Evapotranspiration is the combination of evaporation and transpiration. The first refers to water moving from land surfaces and transforming into vapor, and the second relates to the process of water releasing into the atmosphere from plants. Evapotranspiration is the sum of plant transpiration and evaporation.

† That is, excluding predatory fishing, overfishing, ocean pollution due to the dumping of solid waste (e.g., plastics), or the dumping of untreated effluents (e.g., sewage), etc. These will show up later in the book.

the more CO_2 we emit, the more is available for plants to photosynthe-size and grow. However, bigger plants need to transpire more, which requires more water in the soil. Of course, the consequences increase depending on how much warming we are considering. As usual, the higher the temperature, the harsher the effect. Climate models consid-ering 3°C of warming or more, for instance, show a sharp drop in global agricultural yield.

ENSO: EL NIÑO-SOUTHERN OSCILLATION

You probably know the terms El Niño and La Niña, or at the very least, have heard about them in the news or when a friend is confi-dently explaining why the season has been behaving strangely. We don't have to understand these phenomena in depth for the purposes of this book, but their impacts are so significant that we should at least understand the basics.

Both El Niño and La Niña are phases of the El Niño–Southern Oscillation (often referred to as ENSO). In neutral conditions (which is one of three phases of ENSO), wind blows toward the equator over the Pacific Ocean from east to west (these are called trade winds). Sea temperature is warmer in the west (the region close to Indonesia) and cooler in the east (the region close to Central/South America). La Niña is an accentuation of these conditions, and El Niño is a disruption of these conditions (winds weaken or blow in the counter-direction— west to east). The important thing, however, is not the technical stuff. The important thing is that El Niño and La Niña are parts of cycles that influence extreme weather conditions (heavy rains, droughts, heat waves, etc.). We know they are powerful events with great impacts on our climate, and we actually use our understanding of these phenomena to forecast average weather patterns. However, we're not really sure how human-made climate change will influence these cycles or how the increase in ocean temperature will affect the ENSO phases. There is currently no consensus on how ENSO will alter with climate change or what the consequences will be. Uncertainty when dealing with these powerful climate phenomena is dangerous to play with.

* * *

There is much more to the climate topic than has been covered up to this point, but I'll leave the deeper levels of analysis and discussion of different climate models to the climate experts, as they have a better understanding of the subject and can do a far better job than I ever could. But we did look at the big picture, and within that big picture we dove into some specifics that are important to understand: the increase in the average global temperature, the rise in sea level and ocean acidification, the increase in the likelihood of extreme weather events, and the concerns around our food and water systems, as well as the unknown fate of the El Niño–Southern Oscillation.

However, I would like to call your attention to the fact that until now we have been focusing only on one environmental problem: climate change—or the effects of emitting even more carbon into the atmosphere. But this is only one of the many great environmental challenges we are facing in the twenty-first century and beyond. It obviously has several ramifications, but the core subject of study was climate change. And while climate change is indeed one of the great challenges of the twenty-first century, it would be incorrect to deem it the *only* environmental challenge. But we often do that, don't we? Climate change seems like such an important topic that it often obscures other environmental issues. Not only that, we sometimes assume climate change encompasses other environmental issues because we haven't accurately sorted them out and evaluated where they fit. While there are certainly overlaps between, for instance, climate change and ozone layer depletion, they both have their own place in the field of environmental management and sustainable development. In other words, we could very well solve any ozone depletion problem and still have a climate change problem, or vice versa. There are environmental threats that are directly related to climate change, there are some that are indirectly related, and there are others that are completely independent. The "keep the planet below catastrophic warming" approach is a colossal effort, but it cannot be the only thing on our minds or soon we'll have yet another colossal task creeping up on our doorsteps after we (hopefully) solve this one. Keep in mind the sustainability definition from the Brundtland Report that we discussed earlier and how sustainability goes beyond climate change: "Development that meets the needs of the present without compromising the ability of future generations to meet their own needs."

We conclude our focus in this chapter on the whys and whats of climate change, and we'll move on to a broader view of environmental problems in the next chapter. As we go, we'll see climate change pop up now and again, and I will do my best to relate the new topics we're exploring back to it.

CHAPTER 6

OTHER ENVIRONMENTAL THREATS

YOU'LL RECALL THAT earlier in this book, I mentioned that it would be naïve of me to try to cover all possible environmental consequences of climate change (too many variables, too many interactions, too many uncertainties). As we now expand our scope beyond climate change, this still holds true. This left me with a tough decision to make: Which environmental threats should we cover together in this book? The ones that are currently most impactful? The ones that have the greatest catastrophic potential? The ones that are increasing at the fastest rate? To decide, I turned to a tool that is very helpful in discerning the impact of different scenarios: life cycle assessment (LCA).

LIFE CYCLE ASSESSMENT

Life cycle assessment (also known as life cycle analysis) is a method to evaluate the environmental impact of a product or a service over its (entire) lifetime. It establishes the impact of each step required to manufacture a given product (or deliver a given service) and relates it to the product's use and end of life: Where does it go? Is it recycled, landfilled, incinerated? Better yet, LCA calculates a product's environmental impact in several different categories. This kind of analysis can be useful in product development (e.g., for eco-design), or in later stages, after a product/process has been implemented and you want to calculate its environmental impact or compare it against a different product/process. I've chosen to use the impact categories used in LCA methodology to guide us through the environmental threats in this chapter. Please note that this is by no means the only or even the best ultimate choice, nor is it one that takes into account all possible environmental impacts. But it is a good reference to use, as the impact categories were chosen based on vetted scientific literature. Thus, let us start this chapter by understanding LCA better, so we can then use it to assess the environmental issues that we'll cover, which threaten our lives and the lives of so many other species.

First of all, LCA is amazing! It can answer a lot of questions people have about the environmental impact of their actions and/or their carbon footprint. Which is more environmentally friendly: diesel or petrol (aka gasoline)? Glass or plastic? Paper or plastic? Matches or lighters? Books or a Kindle? Tea or coffee? Cans or bottles? Driving or flying? Cotton or polyester?* I can provide a single answer for all of these in only two words: It depends. But no one likes to hear that answer, so LCA can assist in finding a more specific answer for your particular case, under your particular set of circumstances.

Let's take a look at why it depends, and the many variables LCA has to take into account. In the example of cotton versus polyester, a short-sighted analysis may think that polyester equals plastic, which equals petrol, which equals evil. By the same token, cotton equals plant, which equals good. That chain of reasoning is not incorrect per se, but it is flawed. It forgets that cotton has to come from somewhere; it requires a certain amount of land, water, and fertilizer. It also requires labor and equipment for plowing, farming, harvesting, threading, etc. Depending on where you live, it may not be feasible to fabricate

---

* These questions were all extracted from the Google autocomplete function, so we can assume they are asked often.

cotton fiber for clothing. What if it comes from a place where it is feasible? Great, but we'll have to add the impact of transportation to the equation.

Hopefully you can already see how the answer may change depending on your circumstances, but we can keep going even further. How does the use of clothing made out of cotton compare with clothing made out of polyester? Which gets dirtier after the same use? Does one require more washing than the other? Does the washing process for each have different impacts? And how long does each garment last? (The longer it lasts, the more you can use it without needing to buy another piece to replace it—unless, of course, you buy new outfits regardless of your need.) But here comes another part to this equation, one that is often dismissed: What happens to your garment *after* its useful life? Is it recyclable? Will it be recycled? If it is going to incineration, will it release any toxic compounds? What about the energy it will release—will it be used or wasted? Which releases more energy, cotton or polyester? Oh, it's not going to be incinerated? Where is it going? Landfill? Okay, then will it release hazardous substances over time that will pollute the land, air, or water streams? What about decomposition and emission over time? How long will each stay in the landfill occupying that piece of land?

Are you exhausted with all these questions? You should be. LCA is an exhausting (and exhaustive) exercise. Each of these questions may point you in a different direction when deciding which choice has the greatest environmental impact. So, hopefully, I've convinced you of why the answer is always going to be "it depends" when inquiring about the environmental impact of your choices. In addition to all these questions I've just posed, throw in what was discussed earlier about there being multiple impact categories, and you have a hot mess to figure out. Perhaps the answer to the cotton versus polyester problem for your particular situation is that cotton is better with respect to the global warming impact, but it is worse with respect to the eutrophication impact (we'll cover what this is later in this chapter). LCA is the tool that takes into account all these stages of the life cycle of a product and compiles them into results for different impact categories. It is actually defined by the International Standard Organization (ISO) as precisely this: a "compilation and evaluation of the inputs, outputs and the potential environmental impacts of a product system throughout its life cycle."[78] You can use LCA to discern which product or process is less impactful as part of your personal decision-making process.

In spite of being a great fan of the tool, I won't pretend that it does not have its shortcomings. For starters, it does not deal well with probabilities of events happening (e.g., there is a 30% chance of the cotton needing transportation and a 70% chance of it not needing it) so it is not a good tool for environmental risk assessment. It is also limited by the data available—so while manufacturers, for instance, can create their own processes, we have to adjust our assessment of their product to the information available in existing databases. Here's an example: Recently, my research group and I came up with a new way to recycle solar panels. We can recover several materials from waste panels, such as silver and copper. But our recovery is not of the metal per se, but a mixture of metals and some other materials. The databases available certainly do not contain any information about this particular mixture, but they do contain information about the individual components of the mixture. So I can approximate the environmental benefits of recovering the mixture by combining the benefits of the individual components and creating a "Frankenstein" version with the information available in existing databases.

The database limitation also impedes the use of LCA in analyzing cutting-edge technology. For instance, how will the global shift to cloud computing affect the environment? On one hand, it will require the manufacturing of a great number of electronic components, all of which consume electricity (among other resources). On the other hand, they may very well offset global electricity consumption by means of smarter data processing. Unless there is information on how much savings the new technology can provide (so it can be included in the database), we can't use LCA to assist in this decision-making, at least not yet.[79] Of course, we could run a model based on predictions, but the amount of assumption that goes into an LCA model is already great, and clogging it up with additional assumptions may end up in a result far off from reality, which would have the opposite effect of what we're trying to accomplish.

Here's another example of using LCA methodology to make a choice, which I think you'll find relatable: I recently greatly reduced the amount of meat I eat. This was out of environmental consciousness (which can also be understood as guilt). One of the things I ended up swapping was the toppings of my pizzas. You see, I'd generally go for the "meat lovers" pizzas, perhaps your classic pepperoni or ones with bacon, but when I made the change, I started consuming nonmeat toppings almost exclusively. In the supermarket, I found the best-tasting pizza brand with vegetarian toppings. It was an easy switch and I felt so proud of myself. But at some point, I read the label closely and found out that the brand in question was from Germany—the pizzas were actually coming from Germany all the way to Australia! Did that completely flip the environmental impact equation upside down? Would the environment be better off if I ate a pepperoni pizza made in Australia or a meat-free pizza made in Germany? That is not only an interesting question, but also a solvable one using LCA.

I'm not going to leave you hanging; I actually did quickly run my pizza dilemma through an LCA and found that the veggie pizza still won in terms of CO_2 emission; I was "emitting" less by choosing the veggie pizza that came from across the globe versus the locally sourced meat pizza. But solving this particular problem opened up a whole other world of additional questions: Would it be best to become a vegetarian altogether? And between being a vegetarian and a vegan, which has the best environmental impact? By how much?

This led me to a whole other set of questions that have nothing to do with food: Does it make sense to use a toothbrush made from wood instead of plastic if this means you need to swap toothbrushes three times more often? Would it be best to buy an "eco-product" made of recycled materials that come from afar, or one made of virgin materials made locally?

The questions are infinite and, as you can probably figure out, the answers are completely dependent on the context. LCA assists you in modeling your context and outputting an answer. Moreover, LCA allows us to know whether the answer really matters. You see, it often turns out that we overthink the decisions that have negligible impact on the environment while dismissing the ones that really matter. This will be a consistent theme in later chapters of this book. My point for now is that LCA is a tool that allows us to both compare alternatives and to determine their significance.*

Finally, and this is key, LCA is a tool that assists in *quantifying* environmental impact. This is crucial because it is very hard to improve upon something and track your progress if you cannot measure it. Following up on the examples already provided, maybe a cotton shirt where I live is a "greener" choice, as is eating the vegetarian pizza. But how relevant are these choices? If all consumers were to make the same small choices as mine, would it make a difference? Or perhaps these small choices are so small as to be irrelevant in the grand scheme of things? LCA allows us to answer this. And a good portion of the

---

* If you are interested in running your own life cycle assessments, there are plenty of LCA databases and modelling software available online, both paid and for free. The opensource LCA software openLCA is a good place to start. But you do not need the software to run an LCA. As long as you have the right data, you can calculate environmental impacts with pencil and paper.

third part of this book is dedicated to answering these questions and providing a guide to what we as individuals can do to have the greatest impact we possibly can.

THREATS BEYOND CLIMATE CHANGE

Now that we understand what LCA is and what it's used for, let's move on to the environmental impacts that are generally assessed when doing an LCA. The first is climate change, which we'll skip since the previous few chapters were dedicated to it. Then comes a list of other environmental threats; I will cover the top ones now in no particular order. It is worth mentioning that some of these impacts are given slightly different names in different parts of the world. As an example, abiotic depletion can also be called resource depletion or resource use. It can also be broken down into specific categories, such as fossil fuel depletion and mineral depletion. Generally, the more specific the measurement of the impact is, the more accurate it is. But in this book, we're interested in understanding these impacts from a high-level perspective, so I've chosen the broader categories and the names that are most often used.

Air pollution

As if pumping carbon into the air wasn't bad enough due to its effect on global warming, it also pollutes the very air we breathe, as do many other substances. Air pollution encompasses a lot of topics we have already discussed and blends them all in a soup-like recipe. It is closely related to toxicity (which we will discuss later) in the sense that air pollution is precisely air with a concentration of substances that can cause harm to human health or the environment. The World Health Organization (WHO) estimates that over 90% of people breathe air with more contamination than the recommended guidelines, and that seven million people die yearly due to air pollution.[80] The most prevalent types of air pollution are smog (smoke + fog) and soot. The former relates to gaseous compounds such as carbon monoxide (CO), sulfur dioxide (SO_2), nitrogen oxides (NO_X), volatile organic compounds (VOCs), and ground-level ozone (O_3). The latter, soot, has to do with airborne particulate matter (PM) of chemicals, smoke, dust, allergens, etc. Toxic metals such as lead and mercury can also become airborne and be considered air pollutants.

Particulate matter in the air is classified according to its size, which is indicated by the diameter of the particles (there is a whole field of study dedicated to this topic). Fine particles with diameters smaller than 10 μm (micrometers) are called PM_{10}, and, conversely, $PM_{2.5}$ are particles whose diameter are smaller than 2.5 μm, and $PM_{2.5-10}$ are those with a diameter between 2.5 μm and 10 μm. While PM_{10} includes $PM_{2.5}$ and $PM_{2.5-10}$, they are separated into these different categories because the behavior of the smaller particles in our bodies differs from that of the larger ones (for instance, the smaller particles can eventually find their way into our bloodstream!). Particles larger than 10 μm (up to 500 μm) are also included in the particulate matter category, but they settle relatively quickly and are usually not inhaled. These larger particles are the ones you are probably the most familiar with, and include fly ash,* dust, metals, etc. To assist in building a mental picture: When you make your bed (assuming you do that) and see a beam of light with all those dust particles floating around, those are the larger ones, not the PM_{10} ones. Size matters because it affects how long a particle can stay airborne, which is important when dealing with air pollution and its harmful effects. After all, the longer it remains airborne, the greater the chances we'll inhale it.

_____

* Fly ash is a technical term. It is a by-product of burning pulverized coal in power plants.

The link between air pollution and human health impacts has been well established and is currently widely accepted.[81] However, there is no accepted threshold below which exposure to particulate matter does not cause health issues. What we do know is that both short-term and long-term exposure seem to have harmful effects such as premature mortality and reduced life expectancy. Short-term exposure exacerbates preexisting conditions such as asthma, allergies, and bronchitis, while long-term exposure causes diseases and increases their rate of progression. Air pollution effects can range from minor upper respiratory irritation to severe chronic complications.[82,83] The WHO has stated that millions of premature deaths occur yearly due to household air pollution and linked illnesses such as pneumonia, stroke, ischemic heart disease, chronic obstructive pulmonary disease, and lung cancer.[84]

The effects of smog also range widely. The first effect that generally comes to mind when people hear "air pollution" is the reduction in visibility (but you probably never thought of it that way, just in terms of "smoke" or haze replacing blue skies). But as the compounds react, the formation of other gases responsible for a variety of health effects start to take place. Ozone* at ground level, for instance, can cause lung tissue damage and itchy or burning eyes. But the harmful effects don't just affect humans, as ozone can damage crops, trees, and materials such as textiles, rubber, and some plastics.[85]

But what is the origin of air pollution? Well, the larger particles (greater than PM_{10}) originate from industry, uncontrolled combustion, or power plant boilers (where we burn the fossil fuels to heat up the water/air that will move a turbine, which will power a generator, remember?),[86] but that is not to say that these processes don't also release smaller particles in addition to the larger ones. Another

---

* Remember ozone and its three roles from chapter 4, which we covered when discussing GHGs.

example is internal combustion engines (used in cars, boats, ships, airplanes, trains, and nontransportational equipment such as lawn mowers and chain saws), which are quite inefficient when compared to other forms of energy transformation. In cars, an internal combustion engine roughly converts only a fifth of the chemical energy into motion (kinetic energy).[87] This inefficiency is partially translated into emissions of by-products. Thus, not all carbon from the fossil fuel in question is oxidized to carbon dioxide (CO_2), and a fraction of it is left unburned in the ashes or as soot.[88] In addition to particulate matter and carbon dioxide, mobile sources directly emit methane (CH_4), nitrous oxide (N_2O), carbon monoxide (CO), sulfur dioxide (SO_2), nitrogen oxides (NO_X)*, and the formerly mentioned non-methane VOCs.[89] In the presence of sunlight (which is obviously not much of a limitation in most places), the combination of some of these gases (NO_X, CO, CH_4, nonmethane VOCs) reacts to produce photochemical smog (this is the technical term for the smog we've been discussing so far; "photo" refers to the sun's energy, needed for the reaction to occur). These reactions will eventually produce high levels of ozone, whose effects we have already discussed, and other highly reactive—and harmful—gases.

The sources of the individual gases and particulate matter include mobile sources and fossil fuel power plants, but we can dive deeper and look at the individual pollutants. Sulfur, for instance, is present in different types of coals and crude oils, and its combustion (aka its reaction with oxygen) produces sulfur dioxide (SO_2). It is also present in gasoline and natural gas, though in smaller concentrations than

---

* It's easy to get nitrous oxide (N_2O) and nitrogen oxides (NO_X) confused. The former is the laughing gas used as an anesthetic As we saw earlier, it is very stable and a potent GHG. The latter is written as NO_X because it can take the form of nitric oxide (NO) or dioxide (NO_2), in addition to other nitrogen oxides also present in the atmosphere. NO_X are a family of poisonous, highly reactive gases that contribute to air pollution.

in coals and oils.[90] NO_X gases are produced at high temperatures where there is enough energy to split the N_2 molecules (remember that about 78% of our atmosphere is actually composed of N_2), which subsequently oxidize. NO_X are primarily emitted from mobile sources (again, motor vehicles), but also originate from fossil fuel-burning power plants and industrial activity (e.g., industrial boilers and cement manufacturing), as well as operations specific to nitrogen (nitric acid production and nitrification of organic compounds).[91] Carbon monoxide is another highly reactive gas that, if inhaled, displaces oxygen in the body and in high doses can kill by suffocation (it was actually used as an exterminating agent during the Second World War in some instances![92]). Carbon monoxide, by definition, is produced via incomplete combustion. So we can directly relate many of the sources that emit carbon dioxide back to carbon monoxide (remember how inefficient internal combustion engines were?), but it is also emitted by smoldering coal and uncontrolled incineration of municipal waste.

So far, we've talked about the major sources of outdoor air pollution, which are generally the first ones that come to mind. But indoor air pollution is also as harmful, yet often forgotten. I believe this is because it does not represent the lion's share of emissions, so when compared with the other sources we discussed, it looks small. Nonetheless, because it is so localized (i.e., indoors, and therefore more concentrated and easily inhaled), it is a major health concern. The WHO states that 2.6 billion people (about a third of the population!) still cook using fuels such as coal, wood, and kerosene. Not only are these inefficient, they also release great amounts of soot, causing the health problems previously described. People also use these types of fuels indoors for lighting and heating, which in turn generates indoor air pollution.[93]

Note that we could stop global warming completely and still have an air pollution problem in several parts of the world. It is not hard to imagine such a scenario: We electrify as many processes as we can while also creating a very robust renewable energy infrastructure. Alongside these two important measures, advances in key industries (e.g., the making of steel and cement) allow for a significant reduction in carbon release into the atmosphere. Temperatures peak at about a 1.5°C increase and then slowly come down again. Great! Global warming has been solved! However, a significant number of households still use fossil fuels to cook. Not enough to make temperatures rise again, but enough to harm a good portion of the population.

While climate change and air pollution are distinct problems, they intersect in significant ways. We have already seen that several of the compounds that cause smog are also GHGs, so the greater the air pollution, the greater the amount of GHGs in the atmosphere. But there is another interesting interaction: It turns out that rain naturally "rinses" pollution out of the air. And as we know, climate change will drastically reduce precipitation in some regions. So we can expect higher amounts of pollution in these regions. What about the regions where climate change will increase the amount of rainfall? These will be off the hook, right? I wish I could say yes, but while increased precipitation does indeed assist in reducing air pollution, it may also increase the deposition of certain organic pollutants and the runoff of pesticides.[94]

Terrestrial acidification

When we talk about terrestrial acidification (i.e., acidification of the land, as opposed to that of the ocean, which we saw in the previous chapter), carbon dioxide is not the main villain. Instead, sulfates, phosphates, and nitrates take the stage. The emission of sulfur dioxide (SO_2) comes mainly from the combustion of fossil fuels (yes, the same story again). Nitrates and phosphates are widely used in fertilizers and are also found concentrated in sewage—which eventually find their way to the environment.[95],[96] But a little increase in land acidity won't harm anyone, will it? It very well may. The problem is not humans directly, but rather other species. Plants in particular are very picky when it comes to their preferred (optimum) acidity.[97] And if you've ever studied the food chain (or should I say food web?) you know better than to mess with plants because the problem scales way, way up. The consequences of terrestrial acidification can vary from yellowing of the plant tissue (generally the leaves) to reduction of plant diversity. There is evidence that the change in acidity can spoil seed germination and the growth of roots, which decreases the ability to do photosynthesis (i.e., the plant cannot make as much "food" for itself and therefore will not grow).[98] And if possibly starving a plant to death does not sound bad enough, remember which animal species sits on the very top of the food chain and may end up starving next. In addition, it seems the more we pollute, the worse it gets (no big surprises there). But it also turns out that we have another of those magical positive feedback loop effects because the detrimental effects of terrestrial acidification appear to be accentuated in warmer conditions,[99] and can you guess whose planet is warming up?

Ozone depletion

In the early- to mid-twentieth century, long before it was widely understood that human activity has environmental impacts and that these impacts can have significant consequences, society found great use for a group of gases called chlorofluorocarbons (CFCs for short, but also known as Freon). At first, CFCs seemed great. They had all the properties required to make refrigerants (the fluid that runs up and down your fridges and air conditioners), propellants (used to make products like hair spray or aerosol deodorant shoot straight out of the can), and fire extinguishers (you know, the things we use to put out fires). There was one small caveat though: When these CFC gases found their way into the atmosphere, they would just hang around. Being stable components, CFCs have lifespans in the order of 100 years, which allows them sufficient time to hang out in the troposphere and also diffuse into the stratosphere. If you recall, the troposphere is the closest atmospheric layer to Earth's surface, and the stratosphere is the troposphere's upstairs neighbor. Once in the stratosphere, CFCs start doing their damage.

But before going into why, let us stop and appreciate the stratosphere. This protective layer surrounding Earth's surface has an ozone shield responsible for absorbing most of the ultraviolent radiation* coming from the sun. This radiation can damage DNA, so it's a big threat not only for us humans, but for most forms of life (especially the ones living above water). So you can think of it as planet Earth's natural sunblock and thank it for its daily service of protecting us all. As it turns out, the oh-so-useful CFCs, when entering the stratosphere, interact with the ozone (O_3), producing oxygen molecules (the good old O_2 that we breathe) and chlorine monoxide (ClO), effectively depleting the ozone layer of, well, ozone. This ozone depletion ended up resulting in the famous "hole," which is not really a hole, but rather a region above Antarctica where there is less ozone than normal.[100] People generally have a mental picture of an actual hole, so let's try to come up with a better analogy. Picture an understaffed movie theater. In a normal scenario, only so many people would get through. But the ozone "hole" is like an understaffed ticketing area: It's not that there isn't any staff (ozone), but there aren't enough of them, so more people (radiation) can sneak through without paying for their ticket.

Well, if that wasn't bad enough, researchers found there were other ozone-depleting substances, such as gases containing chlorine (used in

---

* UV is a short wavelength radiation. Building on the discussion around how the greenhouse effect works, you may recall that this radiation "passes through" things easily. This includes your body.

dry cleaning, for instance), or bromine (used as an insecticide for soil fumigation, for instance). Worst yet, most of these are also powerful greenhouse gases! For example, tetrafluoromethane (CF_4) is a product that forms *after* destroying ozone, and it is 6,600 times more potent than CO_2!* So, some molecules are so "evil" that they first destroy the ozone layer and then go on to significantly warm up the atmosphere. CF_4 is so stable that it pretty much stays in the atmosphere forever. Most of the CF_4 is released due to industrial activity instead of reacting with the ozone layer, meaning we fortunately skip the ozone destruction part. But we still get its greenhouse effect. And while CF_4 is a very strong GHG, its contribution is small with respect to the primary GHGs because there is much less of it than the others.[101,102] Much like the CO_{2eq}, there is also the ozone depletion potential, which is measured with respect to the depletion capacity of CFC-11. So we convert the ozone-depleting capacity of all substances into that of CFC-11 to have an easy way to tell which are more powerful.

Luckily (no, scrap that—it wasn't luck, it was very well planned and took the coordinated effort of large groups of people), an international treaty called the Montreal Protocol was signed in 1987 to reduce the use of CFCs and was entered into force two years later. It has since been revised many times (most recently in 2016) and is currently structured not only to stop and phase out the production of CFCs, but also many other ozone-depleting substances. This international example of cooperation has led to a decrease in the size of the "hole." It will still take a while for the atmosphere to heal completely because, as mentioned before, several ozone-depleting substances like

---

* CF_4 is a product of the interaction between the ozone layer and some hydro-fluor molecules, and has a global warming potential on a 100-year time horizon (GWP_{100}) of about 6,600! So *after* reacting with some of the ozone, it actually becomes a GHG with a 6,600 CO2 equivalency. The amount of CF_4 produced in the ozone-depleting event (photochemical process) is small when compared with the amount released due to industrial activity.

to stick around for long periods, but current forecasts predict that the biggest ozone "holes" will recover by 2040. Personally, I find it an especially important treaty because it symbolizes hope, cooperation, and unity. Hope because there was a global problem and we were capable of turning it around; cooperation because it is the most widely ratified treaty in UN history; and unity because of the recognition that the destruction of the ozone layer, regardless of where the actual "hole" is, is a common global responsibility.[103,104,105]

Eutrophication

Eutrophication means the nutrient enrichment of an aquatic environment. This may not sound like a problem at first and, indeed, in certain regions it is not. However, nutrient enrichment can lead to the excessive growth of certain species (e.g., phytoplankton and duckweed). Much like the greenhouse effect, eutrophication is a natural phenomenon and happens in cycles (often related to the seasons). The problem arises when human activities accelerate the rate of eutrophication through the discharge of nutrients into the aquatic environment. The growth of aquatic plants and algae is dependent on several factors such as temperature, carbon dioxide availability (remember that CO_2 is "plant food" used in the photosynthesis process), sunlight exposure, and the presence of various substances such as (bi)carbonates and sulfates. But as far as substance dependence, the availability of nitrogen and phosphorus is the dominant factor. In other words, excessive availability of nitrogen and phosphorus will most likely trigger the overgrowth of these aquatic species. These two main nutrients primarily come from agriculture (fertilizers), industry, and sewage, which eventually find their way into water bodies (aquatic ecosystems).[106,107]

You may ask, *What's so bad about a little more algae in the water?* For starters, it covers up the water surface, allowing less sunlight to get through. That alone kills plants that are deeper in the water and also complicates things for predators that need the light to operate effectively. Moreover, the overpopulation of phytoplankton will consume large amounts of carbon dissolved in the water, which lowers its acidity (this is the reverse effect of the ocean acidification we saw in the previous chapter). If you think this sounds like a good thing, just wait, there's more. In large numbers, these organisms consume large amounts of carbon and release large amounts of oxygen—but this is not a steady state, it is a temporary situation created by the release of nutrients arising from human activity. This population boom is unsustainable, and the resulting competition between organisms creates huge die-off events. Their decay consumes oxygen at massive levels, and that once oxygen-rich region no longer has the very organisms responsible for producing the oxygen in the first place, so the region then becomes hypoxic and anoxic (fancy words to say "with little" or "with no" oxygen), which is deadly for any living creature that depends on it. If killing aquatic life due to nutrient enrichment is not a good enough reason for you to take eutrophication seriously, let me top that off with a few more: degradation of water quality (be it for swimming, drinking, or anything else), destruction of fisheries, and numerous other public health risks.[108,109]

Toxicity

Most people are familiar with the concept of toxicity and understand that substances that can cause harm to humans or the environment must be kept under control. Here is the catch, though: Everything is toxic. It all depends on the quantity of a substance and how it's administered. Have you ever heard the expression "the difference between medicine and the poison is the dose," also phrased as "the dose makes the poison"? Well, it applies to everything. The sodium we ingest as salt is beneficial to our bodies in small quantities, but in excess, it can lead to high blood pressure and eventually kill us. Hey, even water can be viewed like that. In small dosages it is essential to us, but too much can drown us to death or swell our cells beyond repair if ingested. There's also the story of the man who died due to excessive caffeine intake—the equivalent of 200 cups of coffee, to be exact.[110] (Don't worry, you can't die from drinking too much coffee, only from taking caffeine in concentrated pills.) My point is that everything can be toxic if taken in the wrong quantity.

So, when we talk about toxicity, we need to bear in mind that, even in small quantities, some things can be harmful. And "small quantities" is very relative. Breathe a puff of hydrogen cyanide and you are gone; even the tiniest amount is too much for humans. But breathe the same amount of oxygen and it won't be enough; you'll asphyxiate. Thus, when speaking of toxicity, we need to understand what the measurement is in relation to—whether humans or environment (i.e., human toxicity or ecotoxicity)—otherwise known as the "target." Additionally, we need to think of exposure, persistence, and the actual effect of the toxic substance.

Exposure is the contact between the target and the substance for a given boundary (e.g., skin), time, and frequency. Persistence has to do with how long a certain substance will last in a given location. The effect part of the equation takes into account what the substance actually does to its target. It is based on standardized tests that are performed at a given concentration for a particular target (e.g., a certain species). Any harmful consequence can be considered, and it is ranked according to its acuteness. Harmful consequences can range from mortality to reduced mobility, reduced growth or reproduction rate, mutations, behavioral changes, changes in biomass or photosynthesis, etc.

Going back to the "dose makes the poison" saying, scientists use a metric called EC50 to evaluate what dose is "too much." The EC50, also known as half maximal effective concentration, is defined as the concentration of a given substance that causes a measurable effect in 50% of the test organisms under specified conditions.* In humans, the effect impact is relatively straightforward because a single species (albeit with its own nuances) is being considered. But when dealing with ecotoxicity, because there are a myriad of species co-inhabiting an ecosystem, a species sensitivity distribution model† is used to adequately assess the consequences of a particular substance to an entire ecosystem. For ecotoxicity, since it considers a whole environment, the EC50 is further related to the hazardous concentration (HC50), which is the concentration required for 50% of the species to be *exposed* to the EC50.‡[111,112,113]

People tend to associate toxic substances with flashy inorganic elements like lead and mercury. While this perception is not incorrect, it can be misleading. Toxic substances can indeed be inorganic elements or compounds (let's add chlorine gas and hydrofluoric acid as examples to the list), but they also include organic compounds (methyl alcohol, benzene, carbon tetrachloride, and toluene, to name a few) and physical toxins (coal dust, asbestos fiber, silicon dioxide dust, or fine glass particles; recall the air pollution particles we discussed in chapter 6). The latter category is often forgotten but its impacts can be as harmful as any of the others.

---

* Though somewhat counterintuitive, a lower EC50 indicates a more potent substance because it takes a smaller concentration to produce the effect.

† A species sensitivity distribution model first tests the chemical on various species to see what quantity of it causes harm, and then records how sensitive each species is. Graphs are created (called distribution curves) that plot all species studied. The "safe" level of the chemical is usually decided by picking a point on the curve that protects a high percentage of species (like 95% of them) from harm.

‡ You may also see the term LC50, which is the concentration *lethal* to 50% of a given species over a given period of time. So it is similar to the EC50, but instead of considering any effect, it only considers the death of individuals.

But what ruler do we use to measure toxicity? Much like CO_2 is used to measure how potent a given gas is with respect to the greenhouse effect (remember the discussion of CO_2 equivalents from chapter 4), we use 1,4-dichlorobenzene (paradichlorobenzene)[*] as a reference point to be able to run comparisons on toxicity. When dealing with toxicity in life cycle assessments, we generally evaluate the human toxicity potentials (HTP) as 1,4-dichlorobenzene equivalents. Remember how determining the time horizon was important when calculating the magnitude of the CO_{2eq}? When dealing with toxicity, the same idea comes into play. Generally, a longer time horizon (e.g., 100 years or more) is used.[†]

The method does not come without its shortcomings, however. For instance, when dealing with carcinogenic substances (i.e., substances that can cause cancer), one year of life lost for all ages is taken as the unit of measurement. But no future damage to health is accounted for. And here again, uncertainties come into play. For instance, there is still debate over the carcinogenic effect of certain substances to human health. So, the LCA tool splits carcinogenic substances into two groups: the ones with strong evidence of it being carcinogenic and

---

[*] 1,4-dichlorobenzene is a toxic substance that is irritating to the eyes, skin, and respiratory tract. It is considered a carcinogen, and acute inhalation or oral exposure to high concentrations of it may result in liver damage. It is most commonly used as a mold and mildew killer and to control insects, and it's the source of the infamous smell of mothballs.

[†] The choice of time horizon in LCA depends on how far into the future you want to consider the environmental impacts. We generally have three options: (1) Egalitarian, which considers a very long-term time horizon and is focused on the far future (500+ years); (2) Hierarchist, which is the mid-term, balanced option and is focused on the next century (100 years); and (3) Individualist, which considers the short-term and is focused on the immediate future (20 years). Each perspective has its own rationale, and the choice of time horizon can significantly influence the outcomes of an LCA.

the ones where evidence is still insufficient. Yet another uncertainty is around the interaction between substances: Will they simply add their effects? Or will they counter each other? Or perhaps they'll act in synergy and increase each other's power? There is no clear answer, and the addition of effects is commonly considered as a rule of thumb.[114,115]

How is toxicity related to the other consequences of climate change mentioned so far? It is not, at least not directly. But they do feed on each other. Toxic substances can be released deliberately from anthropogenic activity (e.g., transportation, farming, and industry such as paper plants and leather-tanning operations), or as a by-product of the activity. An example of deliberate toxic release is the use of pesticides, while uncontrolled incineration of waste can release toxic dioxins and furans as by-products of the primary process. The consequences of the spread of toxic substances include a decrease in biodiversity, which may spiral down to a weaker biome that is less resilient to change and less capable of absorbing our emissions (less "buffering"). At the same time, there seem to be interactions between a warmer climate and the toxicity of certain substances. For instance, the melting of sea ice releases more seawater into the atmosphere, which may assist in making certain chemicals airborne, such as mercury.[116]

Climate change will significantly impact how chemical pollutants move and behave in the environment by changing the physical, chemical, and biological factors that control their distribution between the air, water, soil, and living organisms. These factors include how chemicals are exchanged between air and surfaces, how they are deposited through rain or dust, and how quickly they break down (such as through sunlight, microbial activity, or reactions in the air). Here, changes in temperature and rainfall will play a major role in how these chemicals spread and settle. Additionally, as temperatures rise, the speed at which these chemicals react in the air may increase, potentially leading to higher toxicity.[117]

So, in general terms, the warmer climate produced by global warming will enhance the toxicity of contaminants, but on the other hand, it is also likely to increase the rates of chemical degradation.[118] There are plenty of variables in these interaction equations. We are talking about hundreds of thousands of substances interacting with a myriad of species, while being subject to varying effects due to climate change. So we may conclude that while there will certainly be interactions between toxicity and climate change, it is still unclear how these will take place on a macro scale and how bad it will be.

Loss of biodiversity (land use)

A whopping 95% of terrestrial lands (excluding Antarctica) have seen human modification to varying degrees, with the remaining 5% being in inaccessible places (e.g., rock and ice within boreal forests).[119] Humanity's ability to modify the environment is nothing short of extraordinary, for better or worse. Land use is tracked as an environmental impact mainly due to the strong correlation it has with loss in biodiversity. A common unit to measure is a given area of land used for one year for a given use (example: m^2·year annual cropland equivalent). This way, we then can track the different types of land use (e.g., agriculture, forestry, urbanization) and convert them to a single unit of reference. The method takes into account the type of land use, the time it takes to back-transform it into natural land, and correlates these with the loss of biodiversity potential.[120]

Climate change is disruptive for us humans for all the reasons discussed so far, but for other species it is generally worse. Thanks to Darwin and Wallace's theory of natural selection, we know that only the species that are better able to adapt to the environment are the ones that survive. However, as Rachel Carson noted in *Silent Spring* (go back to chapter 2 if you need a refresher), human actions are changing the environment so rapidly that many species do not have time to adapt. The time scales of natural evolution and human-caused changes to the environment are completely different. This is especially true with climate change, so much so that some researchers say that these changes have triggered the sixth major extinction event, mainly due to the emission of GHGs, land-use change, and the introduction of non-native species to exotic environments.[121]

Granted, there is a debate among scientists in the field about whether the sixth mass extinction is yet to start or if it has already started. Others say that claims of major extinction are alarmist and biodiversity loss hasn't risen above baseline.* There are a lot of boxes that need to be checked before we can define a given episode as a mass extinction event. But we don't really need to check all the boxes to see what is happening. A recent study assessing the topic mentions that "evidence for a major biodiversity crisis appears overwhelming" and that, although the sixth mass extinction may not have officially started yet, there is certainly an increased rate of extinction and acute loss in populations of species because of human activity.[122] It is always good to remember that the previous five mass extinction events occurred due to natural causes, and this sixth event (if we can call it that) is entirely caused by humans.

Leaving aside the "sixth mass extinction" label, it is crucial to understand that the loss of biodiversity is not an if or a when; it is already happening. We can see it in action today! Early in the twenty-first century, scientists registered the loss of the first mammal whose extinction can be directly traced to climate change. Researchers believe the Bramble Cay melomys (*Melomys rubicola*), a rodent species endemic to a small coral cay in the northern Great Barrier Reef, became extinct because of storm surges and sea level rise, which resulted in vegetation decline.[123]

---

* The argument made by these scientists is that species are always going extinct and what we are observing today is not different from what occurred in other times in history.

The declining population of toad species is often cited as an example of loss in biodiversity due to climate change. Here again, we have one of those situations in which the mechanisms are not neatly established. Climate change induces a myriad of changes, several of which can determine the fate of a given population. Toad embryos, in particular, are generally very vulnerable to variations. Climate change's consequences, such as variability in precipitation and atmospheric contamination (recall our discussions about air pollution and toxicity in previous sections), can lead to a weakening of the amphibian immune system and an increase in the rate of infections. This is only one example of a mechanism that may be happening in parallel to many others.[124]

But what is the overall trend? Is it just some specific species that are going extinct, or can we find a trend to see the big picture? To answer this question, we can turn to the International Union for Conservation of Nature (IUCN), which maintains the most comprehensive database on the extinction risk of animals, fungi, and plants: the IUCN Red List. You can find several well-grounded arguments as to why this list and methods used are not perfect,[125] but it is nonetheless a great source of data. The list currently includes over 40,000 species threatened with extinction,[126] a number that has pretty much quadrupled from the year 2000 as more species are assessed.* Not great news, is it? Also, remember our bias to pay attention to the cute animals and forget about all the others. Plants and invertebrates don't appeal to us like

---

* Watch out for the statistics here! Often the number thrown around is the total percentage of species threatened with extinction. But the percentage is the ratio between the number of species threatened (numerator) and the total number of species assessed by the IUCN (denominator). Since the total number of species assessed has been increasing, the percentage may decrease because fewer species are genuinely becoming threatened and/or because the ratio is artificially deflated by cataloging many species not threatened with extinction. So the absolute number is more important.

polar bears and koalas do, so it is easy to lose track of these creatures even though they tell a very true and urgent story.

In their 2020 report, the World Wide Fund for Nature (generally known as the World Wildlife Fund or by the acronym WWF) identified five main threats to biodiversity.[127] The first is changes to land and sea, which include habitat loss and degradation due to human activities such as unsustainable agriculture, logging, urbanization, and mining. This first threat has been identified as the most significant driver of biodiversity decline.[128] The second threat is the overexploitation of species, which refers to hunting, fishing, or harvesting at rates beyond the rate of replenishment. The third is invasive species and disease, which refers to the introduction of species to a new environment where they thrive due to lack of predators or because they carry new diseases for which the native population does not have appropriate immune systems (the "disease" itself often being a new species—like a virus, for instance). The invasive species can outcompete the native one for resources through unfair advantages. We know that nature tends to balance itself and find an equilibrium, but when a new species is introduced to an ecosystem, nature needs to start to do its thing all over again, pretty much from scratch. There is a good example here in Australia, where a certain species of toad was introduced as a way to control crop pests, but these toads happen to kill the predators who would eat them and keep their population under control (these predators never evolved to develop immunity against a toxin produced by the toad). Now the toad species has become the pest!

The last two main threats identified by the WWF are pollution and climate change. Note how climate change is only one of the five main threats identified, meaning we could completely solve climate change and still have a biodiversity loss problem. Let me stress this: The pathways for extinction are complex and climate change can exacerbate these paths but cannot be considered the sole cause for them.

And like with most global warming-related things, the warmer it gets, the higher the negative effects. The current understanding of exactly how much worse things will get as climate change evolves is poor.[129] What we do know is that limiting the warming is the best course of action to preserve species and minimize biodiversity loss—and that stopping there will not be enough.

Abiotic depletion (resource use)

This category refers to resources that are extracted from the earth for human use. When speaking of *abiotic* resources, we are limiting them to "nonliving" things: metals and fossil fuels, as opposed to wood or crops.* Abiotic resources are viewed from the usage perspective, meaning they are measured based on their usefulness and availability to humans. A copper pipe used in a building will be unavailable to be used elsewhere (or as something else) for several years (probably several decades). In this sense, this impact category ends up being an economic problem of allocation.

Resources can be transformed in a way that they no longer serve as a resource after use, such as natural gas or diesel after they are burned and the energy is extracted. Or resources can be dispersed in such a way that they don't lose value but rather become less accessible for human use, such as a gold bar whose content is used in the manufacturing of millions of electronic devices sold all around the world; the gold is still there but is not concentrated anymore. This is an important distinction; in the latter example, the resource hasn't really been depleted. This is why there are two ways of defining abiotic depletion: in the

---

* We shouldn't be naïve and dismiss things as absolutely "nonliving." Microorganisms are often found in a symbiotic relationship with abiotic resources, making them very much "alive." A good example is coral reefs, which are made up primarily of calcium carbonate (an inorganic "nonliving" compound) that hosts a myriad of species and creates a unique ecosystem.

narrow sense or in the broad sense. The former focuses specifically on primary extraction (think iron ore mines), while the latter considers the resource's presence on the entire planet, including in nature (the biosphere) and in human-made environments (the technosphere or anthroposphere—think of a metal scrapyard containing iron[*]). Iron that was originally in the form of oxide (iron ore), regardless of how it's processed, will never be depleted on Earth, unless of course we are using it to send rockets to outer space.[130]

At this point, you may be thinking, *What does this have to do with environmental impacts?* This is actually an amazing question because the seemingly obvious answer is incorrect. It is easy to create a causal relation between resource extraction and environmental impact. A perfect example is the whole process of mining of minerals, including excavation, beneficiation,[†] and transportation, all the steps with their own significant environmental impact. We can think of the sheer amount of energy required to go from a rock to a metal (e.g., bauxite to aluminum). However—and this is a big however—none of these impacts are accounted for in the resource depletion impact category. Yes, you heard right. None. But they are not completely ignored; rather, they are accounted for in different impact categories (all the ones discussed thus far—greenhouse gas emission, toxicity, acidification, etc.).[131] So what is the environmental impact of abiotic depletion? The answer is...underwhelming. Environmental impacts can be defined as impacts harming human health, biodiversity, and

---

[*] By the way, the scrapyard will technically have steel, not iron. Pure iron is not strong and, therefore, doesn't have many applications. Steel is the solid solution of carbon and iron. But even when iron has been processed into steel, iron is still present and, therefore, still applicable to the broad definition of abiotic depletion.

[†] Beneficiation is the process of improving the quality of an ore mined by removing unwanted materials like dirt, rocks, and other impurities. It is removing the bad stuff from the good stuff.

material welfare, so it is therefore debatable whether abiotic depletion should be considered an environmental impact, as it is fundamentally an economic problem.[132,133] And if you are anything like me, you are now pretty disappointed to find out about this.

If we look at this situation from the perspective of sustainability, however, the argument that abiotic resource depletion should be considered an environmental impact is a bit more convincing. Remember the Brundtland Report from chapter 2? It defines sustainability as meeting the needs of the present generation without compromising the needs of future generations. So if we use up all the copper available today and there is no copper left for the next generation to meet their needs, this would be an unsustainable practice. But the definition of "needs" itself is quite tricky and often subjective—due to the scarcity of resources, attending to a particular "need" will inevitably leave another "need" unfulfilled. The use and allocation of resources will always be an incremental trade-off.[134]

Yet another argument for considering abiotic depletion an environmental impact has to do with the lengths we, as a species, go to obtain a given resource. Remember all the environmental impacts associated with resource extraction that are considered by other impact categories (e.g., fuel used in mining, energy required in refining, emissions from reduction processes)? Well, these impacts grow as the concentration of the minerals in question decrease.[135] A simple example to illustrate this point: If you are extracting silver in a mine in which the ore grade is 800 grams per ton, this means that for every ton of "land" you extract and process, you'll end up obtaining 800 grams of silver (though in reality less because your process is never 100% efficient). If these mines are now depleted and you only have the option of mining in places where the grade is 8 grams per ton...well, you need to move 100 times more land for the same amount of silver. Of course, technology breakthroughs can happen in the interim and greatly increase efficiency, but as a general rule, the less available a certain abiotic resource, the more energy and resource intensive the process to obtain it will be, which brings about an increase in familiar environmental impacts. In this sense, the depletion of abiotic resources is indirectly related to environmental impacts.

A good example of this in action is sand; it may surprise you to learn it is one of the most used resources by humans. We extract more of it than we extract oil and wood! It goes mostly unnoticed by the general public, but think about it: Glass, concrete, electronics, and solar panels all require sand as raw material input. Glass is mostly sand (silicon dioxide, or SiO_2); concrete is a mix of sand (~30%) with other things; electronics are based on using silicon as a semiconductor, which is pretty much highly purified sand; and the vast majority of solar panels are made of ultrapure silicon—again, highly purified sand. Ideal sand for construction is extracted from freshwater sources like rivers and lakes, and this activity often leaves damaged ecosystems behind.[*][136] And guess what? The more we extract, the rarer the resource becomes, the more extraction is necessary, the more damage, and so on and so forth. Yet another positive feedback loop.

Speaking of solar panels and abiotic depletion, if I asked you, "What is more environmentally friendly, electricity from solar panels or from coal?", I would bet you would answer, "Solar panels." And you'd be right! Well...for the most part. Remember that the answer is always "It depends." In this case, the answer is that it depends on which impact category is being compared. For most categories, yes, electricity from solar panels is better for the environment than electricity from coal power (because of climate change, human toxicity, and many other impacts), but for abiotic resource depletion, the answer is the exact opposite. The manufacturing of solar panels requires rare materials such as silver. Coal, on the other hand, in spite of all the extraction to

* Again, these damages are calculated in other categories. Abiotic depletion only measures the use of sand and its reduced availability thereafter.

date, is still abundant. So, if we look through the single lens of mineral depletion, solar panels are worse for the environment. But I'll repeat myself so you don't close this book with the wrong idea: Solar panels are better for the environment than coal power plants overall and for the vast majority of impacts we measure! In this case, abiotic depletion is the exception, not the rule.

Interestingly, the "depletion" part of this impact category is also not that easy to define and interpret. If we take the case of fishing grounds (yes, I know fish are not abiotic, but just bear with me) to determine the rate of depletion, we first measure the amount of fish ("reserve"), then consider both the rate of de-accumulation (in this case, rate of fishing) and the rate of regeneration (fish getting married and having baby fish). Similar considerations can be made for an entire ecosystem. The same is true for abiotic resources, but in this case, the rate of regeneration is considered nil because, using the example of fossil fuels, it takes literally millions of years for Earth to make the stuff, and unfortunately gold particles don't get together and make baby gold, or else I'd be rich.[137] Problems arise, though, when measuring the amount of resources available. It has long been noted that deposits of a given resource are not fixed and must be continuously updated and reassessed in the light of new geologic knowledge, new technologies, and changes in economic and political conditions.[138] Therefore, much like the case of strong and weak sustainability discussed in chapter 2, defining depletion will always be impossible to a degree, since it depends on the future availability, future technology, and future demands, all of which cannot be tested empirically in the present.

Overpopulation?

What is the trade-off between sustainability and population growth? Is it even possible to have environmental sustainability along with population growth? According to the bad guy in *Inferno*, one of Dan

Brown's many bestselling thrillers, the answer is no. Let me summarize the plot for you: The protagonist needs to stop a plague that was bioengineered to target humanity in order to avoid overpopulation. The villain is a wealthy genetic engineer who owns a big corporation and tries to warn the authorities about the risk of overpopulation. He shows a graph indicating the exponential population growth and projects that it is unsustainable to maintain such a trend without the suffering of most. According to the bad guy:

> "Animal species are going extinct at a precipitously accelerated rate. The demand for dwindling natural resources is skyrocketing. Clean water is harder and harder to come by. By any biological gauge, our species has exceeded our sustainable numbers.... Under the stress of overpopulation, those who have never considered stealing will become thieves to feed their families. Those who have never considered killing will kill to provide for their young. All of Dante's deadly sins—greed, gluttony, treachery, murder, and the rest—will begin percolating...rising up to the surface of humanity, amplified by our evaporating comforts."[139]

This argument is admittedly convincing, and if you accept it, you may end up rooting for the bad guy. But does the argument hold up in the real world? Well, no. A common mistake people make when evaluating trends is zooming in on a time period that is either too long or not long enough. If we take population growth over the last few decades and expect that to be a continuous growth rate, we'll indeed find that we're headed for overpopulation (by definition, *over*population implies something that is *not* sustainable). Likewise, if we took the economic growth rate of the decades following the Second World War and expected that to be continuous, we'd be horrified by the actual (much lower) growth rate that was measured in recent years (as was the case with many heads of state and economists throwing blame around in the 1980s and 2000s). In his book *Capital in the Twenty-First Century*, French economist Thomas Piketty makes a case to show that the accelerated growth rate after World War II was the exception, not the rule. In reality, growth of most things (especially living things) tends to follow an S-shaped curve. And it turns out our brains are not very good at grasping S-shaped curves. Think of the height of a baby, then the height of a 10-year-old child, then the height of a teenager. If we take this rate of growth and apply it as a continuous trend, we'd have 40-meter-tall (130-foot-tall) adults at the age of 60. But that does not happen. Instead, human beings grow fast during our younger ages and then growth plateaus when we are older.

So what is the expected trend in population growth? Like all projections, answering the question of what the world's population will be in the future has some uncertainty. Because of this, the UN has three different scenarios, none of which are doomlike. The simple explanation to why overpopulation is not a concern is this: The more economically developed a country is, the lower its fertility rate. In other words, the richer a country, the fewer children per family. Another way to look at this phenomenon (which is closer to the actual cause-and-effect relation) is that countries with the highest mortality rates also have the highest population growth. In communities where children survive and are not required for labor, where women get an education and participate in the labor force, and where there is accessible family planning, the number of children per woman drops significantly.[140] At the time of writing, we are at roughly eight billion human beings worldwide. The UN's expectation is that we will peak at about 11 billion.[*][141] Eleven billion sounds like a magic number, but the expla-

---

[*] The 2100 projection bracket is 10.9 billion (9.4–12.7) with a 95% prediction interval, which is the range within which a future value is expected to fall given the uncertainty in the data and the model. The most recent projections actually point to 10.4 billion.

nation could not be more straightforward. Think of a family that has two children. If these children have two children and so on, there is no population growth, right? The parents eventually die and leave two children to take their place, and the loop goes on. Well, that is kind of the rule today already, as the global average fertility rate is roughly two children per woman. So where do the additional three billion come from? It's just the parents getting older! They're simply hanging around in the population for far longer than they used to.

Take the fact that most countries with high income* have two or less children per family. These countries cannot expect their population to grow "naturally" (migration could still increase the national population, but this would not contribute to the global population). In general terms, these countries also have older populations; that is, they have higher life expectancy, which means they have people evenly occupying all age brackets. Then consider the other extreme, in countries where the fertility rate is very high and life expectancy is not as high. They have a "gap" in the older age brackets. And it is precisely this age gap that will be occupied more in future years as life expectancy is anticipated to rise in these countries. There are approximately two billion people in each major global age bracket (0–14, 15–29, 30–44, and 45+) at the time of writing. Now let's split the 45+ bracket into three additional brackets: 45–59, 60–74, and 75+. These are the older age brackets to be filled. There you go: three brackets, one billion more for each, three billion additional humans.[142,143]

Another recent study revisited the question of population growth and predicted that the world population is projected to peak at around

---

* This "high" income would not even be considered high for most people reading this book. The two-children-per-woman average is a reality for countries with a per capita GDP of $12,000 USD or higher, which includes countries like Brazil, Vietnam, Peru, South Africa, and Azerbaijan, to cite a few.

10 billion people in 2064 and then *decrease* to roughly 9 billion in 2100. This involves populations decreasing in several countries (some by as much as 50%!).[144] As mentioned, these are projections, so there will be uncertainty, but I find the projection of 9 to 11 billion people to be quite convincing. Please don't get me wrong; there is a huge difference between 9 billion and 11 billion people, and if we consider the confidence interval of these projections, the range is even higher. But the main point remains: The majority of credible projections show that the global population should reach a plateau.

This data allows us to slide into an important and straightforward conclusion: If we expect to peak at ~11 billion people, we need to plan our resources so they will be sufficient for ~11 billion people. We need to arrive at a lifestyle (resources consumed per person, waste generated, amount of GHGs emitted per person, etc.) that will accommodate ~11 billion human beings in the long run.

There is, however, a caveat worth mentioning. These projections assume that, in spite of the anticipated increase in life expectancy, all people will eventually age and die. In his book *Lifespan: Why We Age—and Why We Don't Have To*, geneticist David Sinclair argues that age can be viewed as a disease and combated as such. We might soon see scientific progress to a point where lifespan is greatly increased, perhaps indefinitely. While there are plenty of ifs and buts in such a speculation, it is nonetheless worthwhile to ponder such not-that-futuristic scenarios and think about their consequences to the environment and homo sapiens' relationship with them. Sinclair argues that even if the life expectancy is boosted by only 10 years, this would have such a significant impact that it could take the natural balance to a breaking point. If this is the case, should we toss the ideas in this book away and completely rethink human-environment interaction? Well, yes...and no. Indeed, when we account for the possibility of a significant increase in lifespan or in the complete elimination of aging, our priorities and problems will shift, but not only in the environmental sphere. They'll change the shape of society as a whole. Our political systems, economic systems, family interactions, moral compass—the changes would be enormous. Even if several ideas in this book become obsolete, the laws of science that constrain reality will be conserved. The same amount of energy will reach planet Earth from the sun, the greenhouse gas effect will still be present and measurable, and carbon atoms will still have six protons and six electrons. Thus, the ideas around what can be done on the individual and collective level will still be valid, albeit possibly not enough. So, no. Don't worry, you're not wasting your time reading these pages, even if your time does become infinite, eventually.

* * *

Just as it would be naïve to try to write about all the consequences of climate change, so too would it be to try to cover all the additional environmental impacts. But I hope that this brief summary, and the introduction of the life cycle assessment tool, has given you a sufficient big-picture view of some of the issues we face as a species and as a planet, and how to measure the impact of our actions.

We now understand that the terms global warming and climate change were both created to mean the same thing, but the latter is more accurate. This is because the consequences are way greater than an increase in temperature. We also covered other environmental impacts related to human action that are not necessarily related to climate change. These are based on life cycle assessment studies, a powerful tool that you now appreciate a bit better. Remember that this tool exists if you ever need to evaluate a specific product, plan, or situation with quantitative data. It is also useful when you want to challenge the claims of a new initiative—you can cut through the noise of biased information by saying, "Show me the life cycle assessment of this endeavor."

There are certainly other problematic consequences arising from the way we consume and dispose of our resources—from antidepressants and painkillers that are excreted by our bodies and end up in crustaceans in the Arctic,[145] to the fertilizers used to grow the food we eat that end up in remote water bodies and cause all sorts of changes to a given environment. Much like the silent and far-reaching consequences of DDT use that were studied and explained by Rachel Carson, there are several other changes we are constantly imposing onto our environment, creating unknown consequences that we have yet to study better. We ought to listen to scientists and take them seriously when these consequences come along.

This chapter—in addition to the previous chapters about climate change—has hopefully shown that there is so much more than we can fathom individually, and that we require researchers and activists to fight the various ongoing battles to protect the environment in its many nuances. Nature is beautiful and has found a way to reach equilibrium in each little ecosystem we can think of. But we humans are good at messing with such equilibrium, and I don't see us stopping any time in the near future. Many of our actions have proven to be extraordinarily harmful, so much so that they endanger the future of our very species. But as a collective species, we don't seem to be too worried, do we? If it really is that bad, why aren't more people paying attention?

Well, for starters, consequences of environmental changes are generally felt by those who are most vulnerable. It will take a while until most people, especially those in high-income countries, actually feel the negative consequences in their own communities and in their own skin in a loud and clear manner. But the most vulnerable are already suffering and, in some cases, being exterminated by such consequences. This is easy to spot when we look at vulnerable ecosystems and species that have already become extinct, but the devastation is also being felt by human beings around the world. We are bombarded by stories of communities ravaged by increasingly extreme weather. Severe drought, flooding, and wildfires are becoming more frequent. According to the UN's refugee agency, between 2008 and 2015 over 200 million people were forcibly displaced from their home countries by weather-related events.[146] This number has only grown since then.

Still, if you are reading this book, chances are you haven't felt much significant change in your life, though you may have noticed increasing extreme weather in your area, more days of bad air quality, or perhaps your communities (and politics) are feeling pressure from the influx of climate refugees. Hopefully, you—and the rest of us—will take action before you start to feel the effects more acutely.

Keep in mind the famous Holocaust poem often attributed to Martin Niemöller that says silence in the face of oppression enables injustice to grow. The poem highlights the danger of not speaking out against persecution, even when it doesn't directly affect us, because ultimately, when we remain silent, there may be no one left to defend us when we become the target.

Climate change first came for the most fragile species, like the small Bramble Cay melomys rodent. It then came for the most vulnerable people: Low-income countries and small-island developing states endure the harshest health impacts from climate change, despite their minimal contribution to it.[147] It already came for thousands of people during the European summer of 2003, Australia's Black Summer of 2019–2020, and the rampant wildfires across Canada and the USA throughout the 2020s and more recently in Brazil and Bolivia,[148] not to mention the increasingly catastrophic hurricanes around the globe. And even when natural disasters that are not related to climate change or any of the threats we talked about take place, such as the devastating earthquake in Syria in 2023, climate change exacerbates the problems and consequences, since severe droughts and high temperatures affect water and food availability, among many other issues.

If we don't act while the changes are coming for others, it will be too late when it comes for us.

PART III

THE SOLUTIONS

CHAPTER 7

LITTLE STEPS: INDIVIDUAL ACTIONS WE CAN TAKE

A FTER READING THROUGH the previous chapters, which illustrate the various harmful consequences of climate change and other environmental impacts, it would be easy to feel discouraged, even angry. Why don't we act? Why isn't this the main topic every day in our casual conversations? Why isn't it the top headline in every major media outlet?

I think the answer, at least for those of us who care, has to do with how our brains are wired. The cognitive biases theorized by cognitive scientist Daniel Kahneman and discussed by George Marshall in his book *Don't Even Think About It: Why Our Brains Are Wired to Ignore Climate Change* seem to offer a very plausible explanation as to why more people are not concerned about the issue. According to Marshall, the three major biases in question are related to loss aversion, temporal discounting, and certainty confidence.*[149] Climate change appears to

---

* Loss aversion is when people fear losses more than they value equivalent gains; for example, losing $100 feels worse than gaining $100 feels good. Temporal discounting is when people tend to undervalue or ignore future rewards or consequences because they seem too far away, favoring immediate gratification instead. Certainty confidence is when people are more confident and feel better when outcomes are certain, even if the certainty doesn't necessarily lead to better results.

be perfectly aligned on all sides of our human biases in a way that compels us to ignore it.[150]

First, environmental issues are always framed in terms of losses: the disasters we may witness, the degradation of our natural environment, the loss of ecosystems. But these losses don't feel very concrete, do they? When we generally think about environmental disasters, we think of either grandiose changes—big fires, big droughts, big floods— or abstract, long-term concepts like "extinction" and "melting ice caps" that are difficult to conceptualize in a tangible way (when was the last time you experienced a biblical flood?*). We don't really picture specific losses of our possessions or things we use and need in our everyday lives. When we cannot picture—or perhaps more importantly, *feel*—the potential losses with precision, it is difficult to understand them and act accordingly.

Conversely, it is very concrete and easy to understand short-term losses: We want the convenience of driving a car and are not willing to give it up. We want a piece of meat and not having it will make us grumpy. We want the cheapest option because buying the more expensive green option will leave us financially worse off. This thinking also applies to the upper corporate and government levels too—sustainability initiatives and green business practices will hurt the bottom line *now*; shareholders and corporations will lose profit

---

* Update: After I finished writing this book and was in the process of publishing it, a major biblical flood hit my home state in Brazil. Hundreds of thousands of people lost their homes and were displaced. Cities completely disappeared in a matter of days. That same year, in North Carolina, where my publisher is based, Hurricane Helene devastated a mountain area that was considered by many experts to be relatively "safe" from climate change. Entire towns were washed away and many lives were lost. So, while this was supposed to be a humorous passage about how the severe weather events are not always obvious, we now seem to be beyond that. The events are obvious. And immensely tragic.

now; voters (or, more likely, powerful political donors) will be mad about environmental regulations *now.* The human brain is wired to be more sensitive to short-term costs and benefits than long-term ones. And what is more long-term than climate change? The need to make sacrifices now for the benefit of future generations doesn't work well with the instant gratification instincts of homo sapiens.

People are also biased toward reacting to certainty while ignoring uncertainty. There is an avalanche of experiments run by psychologists showing this. How does climate change fit in? Well, it is a problem surrounded by uncertainty because of the simple fact that there is a myriad of complex interactions that all affect one another, which may change with every new piece of evidence collected—such is the nature of science. Some of these threats, such as loss in biodiversity, are not only less likely to make the news, but are also harder to track and measure as they involve a very complex web of interactions far beyond the change in climate. This makes a perfect combination of existential threats that are difficult to comprehend based on the way our human brains have evolved to think.

Despite our cognitive limitations, humans have proved to be a most resilient and adaptive species. We have survived countless plagues, natural disasters, ice ages, and more over the course of our species' history. The examples of environmental activism and progress in chapter 2 are a testament to the power of individual, collective, and government action—which is the focus of the next part of the book. We'll go into how this works on the individual level, what the latest data says, and where you can have the greatest positive environmental impact. I'll also debunk some myths and analyze situations that are often portrayed as eco-friendly, when in reality they do little to nothing. The good news is that the opposite is also true: Actions that are often disregarded can have significant impacts.

This chapter is meant to empower you as an individual to take action, to highlight what exactly the role of the individual is, and to show where and how to have the biggest impact to make your effort count. This is important for a couple of reasons. First, to ensure that people with the will to act are not doing so in vain, neither witnessing their goodwill fade away or, even worse (and quite common), taking action that hinders the very objective being pursued. Second, as argued by George Marshall in *Don't Even Think About It* and echoed by Michael E. Mann in his book *The New Climate War: The Fight to Take Back Our Planet*, people will take little action if they believe climate change is an unavoidable condition, but they will be willing to rise to the challenge if it is presented as an active and informed choice and one that brings a sense of shared responsibility and social belonging.[151,152]

I can summarize the whole of part III with three key phrases: (1) be more informed, (2) promote change, and (3) teach. By reading this book, you are already tackling number one. Understanding the problem and the solutions is crucial, especially in an era of misinformation and greenwashing. People are keen to help the environmental cause more than ever before, which is amazing. But it opens the door for those who want to use the cause to push their own products, services, and ideals—and for those who want to gaslight us into believing that our individual actions bear the primary responsibility for both causing and fixing our planet's environmental problems.

Promoting change is the hardest to both grasp and do because it is difficult to measure on an individual basis, and it is much more complicated than it seems at first glance. Yes, we want to ensure our individual ways of living have a minimum environmental impact, and we should strive for that. But I have some bad news for you: That is simply not enough. If we just play the role of responsible consumer, even with the best of intentions and actions, it will still not be sufficient to solve the existential threats we face as a species and as a planet. What we really need are the big steps of structural change, which is what we'll discuss in more depth in the next chapter. The unfortunate truth is that even if we all did everything I outline in this chapter, it would barely make a dent in the devastating harmful impacts our species has on our planet. I do not say this to discourage you, but to provide a much-needed reality check.

This chapter focuses on the little steps; that is, what we can do as individuals to live more environmentally sustainable lives. While these actions alone will not solve our climate crisis or other looming environmental threats, I believe these little steps are necessary for building the kind of collective momentum we need to take on the bigger steps of systemic change. It is only through educating ourselves, taking action, and being examples for others that we will create the massive cultural shift required to make the more transformative ideas in chapter 8 a possibility.

Is there light at the end of the tunnel? Can we still save ourselves and our planet? Yes, we can. This chapter is about the first steps we can take to get there.

CHALLENGE MISCONCEPTIONS TO BETTER UNDERSTAND THE WORLD

Simply learning more about the world is a powerful but often overlooked force for change. Opening your awareness of different cultures—about both their beauty and the challenges they face—and of population density worldwide and international statistics—can assist in increasing your empathy and combating prejudice. Our prejudices are strengthened by the unknown, and it is hard for humans to care about the struggles of another culture if we cannot find ways to relate to them. We are isolated by our alienation of other people, making it easier to ignore their suffering. We forget that we are a single human species, that what happens to the Earth happens to all of us.

My first recommendation to challenge your misconceptions and broaden your understanding is to read a book called *Factfulness: Ten*

Reasons We're Wrong About the World—and Why Things Are Better Than You Think by Hans Rosling and his son and daughter-in-law, Ola and Anna. As the title suggests, the book explores common yet significant misconceptions people generally have about the world we live in. For instance, the world currently has about eight billion people, but would you be able to place those eight units on their correct continent? In other words, how many people would you place in the Americas? What about Africa? One billion, two billion, three billion? Most people get this answer wrong and split it incorrectly. Hell, I got it wrong the first time.* Don't you think it's significant that most of us geographically misplace, oh, *one billion* human beings? That is roughly 12% of the world population!

In addition to the book, the authors also run the Gapminder Foundation, which is dedicated to tackling ignorance about the world and exposing such ignorance in an engaging and entertaining manner. They show that ignorance is not limited to a certain social class or group, and it can be found across the board in different countries, levels of education, and professions. They found that among policymakers and media professionals (the very people for whom such ignorance could be disastrous when they are doing their job), the measured awareness was not significantly different than other groups. You'd think the people deciding where resources, economic treaties, and restrictions are put in place would know roughly how the population is distributed worldwide, wouldn't you?

In addition to their battle to fight ignorance worldwide, the Gapminder Foundation also has a powerful tool called Dollar Street that can deeply enhance your understanding of the world. Using an

---

* If you want to check your guess, here's the current global distribution of Earth's eight billion humans: one billion in the Americas (including North, South, and Central), one billion in Europe, 1.5 billion in Africa, and 4.5 billion in Asia.

interactive online tool, they condense the entire population of human beings onto a single street with households organized by income level; the people with the smallest income live on one end of the street while people with the highest income live on the other end, with everyone else in between. What would such a street look like? How would people of similar incomes in Brazil, Indonesia, Belgium, or Zimbabwe live? How does life differ for people on different ends of the spectrum? Dollar Street uses photos and videos provided by a network of collaborators that illustrate the lives of families going about their daily lives, showing their bathrooms, cutlery, toothbrushes (or lack thereof), footwear, house, and other quality of life markers.

Dollar Street was a striking discovery for me. Actually *seeing* the world and the people living in it showed me that humans are indeed very similar to one another regardless of their geographical position, and that our cultural differences, while apparent, are mild and particular to one thing or another (for instance, the type of toilet used). At the same time, it shows how the things we generally think of as being different are due to the variation in income—the type of house you live in; the state of your teeth, feet, and hands; the utensils you use. Dollar Street shows that the real world is not the one we have created in our heads using the small and insufficient sample of our own lives.

The last thing Dollar Street did for me was to show where I really ranked in the world income chart. It is not only me, though. The vast majority of people with whom I have been engaged in my life (friends, family, colleagues, teachers, students) are on the very right edge of the street (the high-income side). And chances are, the situation is the same for you, too. I've long known how privileged I am and how lucky I've been to have all my needs met since I was a little kid. Intellectually, I've understood the mathematical implications of my income bracket. But actually *seeing* what it means to have such an income—what it allows me to have and do, and how that compares to the life experiences of other people around the world—is a completely different perspective. We tend to assume we're located somewhere with evenly balanced sides, because we know there are people poorer than us and people wealthier than us. I knew I was somewhere on the right half of the spectrum, but I didn't realize how far away from the center I actually am.

Go see for yourself. If you are comfortably reading this book in your spare time, I promise you, you are nowhere near the center of the global wealth spectrum. Understanding how the rest of the world lives will help you put the world in perspective—and, hopefully, help you care about it even more. It will also help you understand that environmental regulations should not be enforced at the expense of the economic development of the countries that need it the most (in this case, the countries whose populations are largely concentrated on the left side of Dollar Street).* This idea will show up in more detail in later pages.

OUR CARBON FOOTPRINTS

Each of us, through our lifestyle, consumption, and daily actions, has a certain environmental impact. A good place to start thinking about environmental impact on the individual level is by looking at your carbon footprint. Understanding where you are on a polluter scale is the first step toward doing better and knowing where to focus your efforts. There are several websites that can assist you with this. One that is straightforward and quite ludic is footprintcalculator.org. It asks you a bunch of questions about your lifestyle (your diet, transportation habits, house size, etc.) to give you a general idea of your footprint.

---

* I find it important to note that the Gapminder Foundation also combats the idea of splitting the world into "countries with poor people" and "countries with rich people." The world today is much more diversly distributed, with countries having the whole income spectrum.

It is not very precise (don't take the numbers too literally), but it is enough to find your general place on the polluter scale. It also calculates how many Earths would be needed to maintain your lifestyle if everyone were like you. So, if your result displays "4 Earths," we would need three additional planets if your habits were universal. Talk about pressure, huh? My guess is that if you live in an affluent country, you'll most likely "require" more than one Earth to satisfy your consumption level. This tool also helps us reflect on the impacts of others around the world. How do you think people from other places (or on different parts of Dollar Street) would differ from you in their answers? Why is that? Where and how do your lifestyles diverge? How many Earths do you think we currently need if we take into account the diverse lifestyles of every human on our planet?

How much can we use without using too much? This question always gets me thinking. For instance, picture a river. A deep, large, beautiful river. You hike to find the river, and once you get there, you drink some of its water to quench your thirst. Is the amount of water removed detrimental to the ecosystem? Probably not, right? But now let's say you bring some bottles to fill up. Still no real impact. Then you decide to divert part of the river to a nearby corn plantation so you can have some natural irrigation going on. Now have you affected the ecosystem? By how much? To the point where it can't recover? In other words, have you stressed it beyond its natural resilience point? The same can be asked of a forest. We find ourselves in a bushland and grab a couple of fallen trees to make a fire. Probably no big deal. Sure, the trees, although dead, would still contribute to the ecosystem of the forest, but taking some probably won't damage it permanently.* Which brings us back to the question: How much is too much?

---

* "Permanently" is being used loosely here. Given enough time, nature can recover (but perhaps not with Homo sapiens being a part of it).

Because we are generally biased to condense our understanding of the whole world to only the world we see around us, it is tricky to envision reality as it is. To assume your answer to the "how many Earths would your lifestyle require" question is equally applicable to everybody on Earth would be incredibly biased.[*] According to foot-printcalculator.org, as of 2020, we need approximately 1.6 Earths to sustain all of humanity's current rate of consumption on average. Thus, we would need Earth to be 60% bigger than it currently is, or we'd need a supplementary planet about half the size of the Earth to provide the resources we use and absorb the waste we produce—in other words, to keep within Earth's natural biocapacity.[†] A different (and clever) way to answer the question of how many Earths we currently need on average is synthesized in the "Earth Overshoot Day." The idea is simple: The Earth (that is, nature) is resilient and therefore capable of replenishing resources at a certain pace. As long as we take and use these resources at a rate below that of replenishment, equilibrium is maintained, and we don't have to worry about damaging the environment.[‡] However,

---

[*] A very good place to correct your biases and to destroy prejudices you did not even know you had is the Gapminder website, mentioned earlier. You'll see that "rich people" (which probably includes you, by the way) live quite similar lives, regardless of where they live. And people struggling share mostly the same struggles all around the world.

[†] Biocapacity refers to the ability of the natural environment to regenerate what people use and to absorb (all types of) waste. It essentially measures the availability of natural resources like clean water, air, and land for agriculture, as well as its capacity to handle the waste we produce. When we use more than the Earth can regenerate, it leads to a deficit, reducing its biocapacity over time.

[‡] The reality is a bit more nuanced. One can take less than what nature can replenish and still drive a species to extinction, for instance. Thus, the overshoot day is by no means the ultimate way to measure how well we're doing, but it does a good initial job at making an individual reflect on their own impacts.

if we take too much, we'll be stressing the environment beyond its ability to recover. So, by taking the rate at which nature is capable of recovering and comparing it to the *average* of human habits, we can arrive at the overshoot day. In other words, the Earth Overshoot Day marks the day of the year in which, from that day onward, we are taking too much in that given year. In 2020, for example, August 22 was Earth's Overshoot Day. From August 23 onward, we were taking more than Earth could naturally replenish. Put differently, from the end of August until December 31, the resources extracted, paired with the environmental damage (emissions, pollution, contamination), pushed beyond the planet's natural resilience. Of course, the ideal scenario would be to not have an overshoot day at all. Or, in the best worst-case scenario, to have it on December 31 every year (playing with fire, right on the edge).

The Earth Overshoot Day has been arriving sooner each year, an increasingly worrying trend. Granted, there have been years in which the date reversed the trend and arrived later (which is a good thing), but overall, the date has moved from December all the way to July since tracking began in 1971. Interestingly, the global pandemic of 2020 produced the first significant reversal, the only one we have seen in about a decade. The reasons are yet to be comprehensively studied and completely understood, but one can reason that the slowdown in economic activity, road transport, and air travel were major contributors. This shift obviously came at a high cost, and no one in their right mind would advocate for a pandemic-like scenario to fight our climate problems. What the pandemic did showcase, however, was that a correction is possible. It sparked hope that we can turn things around. It's as if you are trying to push a huge object and no matter what you do, it does not seem to move. After years trying to push it, you start to feel like it can't be done. But then something happens, maybe the wind blows a certain way, and your push moves the object ever so slightly. Now you've seen it can be moved, and your willpower has taken a shift in perspective.

It's no mystery that not all countries pollute alike. We can make a clear distinction between countries that have a biocapacity deficit and those that have a biocapacity reserve. As a rule of thumb, small but rich countries display great deficits; that is, their pattern of resource consumption and waste production (here, waste includes emissions, solid waste, effluents, etc.) is several times greater than the capacity the country is capable of undertaking resiliently. Examples are Israel, United Arab Emirates, and Singapore. On the other end of the spectrum (and again, as a rule of thumb) we have countries rich in natural resources, whose general population lives within humble means when compared to affluent countries. Examples are French Guiana, Bolivia, and Gabon. The big players that have the biggest impact on planet Earth when it comes to environmental impacts are China, the USA, India, etc.—countries with significant populations and economic activities. But because of their land size, which plays into their biocapacity, these countries end up behind small countries like Singapore when ranking countries by their biocapacity deficit.

There are so many other ways we can classify nations on the environmental issue scale. And you can probably imagine that these different classifications are used to favor the side of whoever is arguing for this or that. If we classify the highest polluting countries in absolute terms, we'll have China, the USA, and India at the very top. But the population gap between the USA and the other two is clearly enormous. So enormous that if we look at the highest polluting countries based on emissions per person, the USA still ranks high, but China and India no longer lead. Instead, Canada, Australia, and additional low-population countries (Qatar, Luxemburg, and Trinidad and Tobago) take their place at the top of the scale.

One pattern that is quite clear is that the higher a country's Human Development Index,* the higher its consumption-based CO_{2eq} emissions. When we think about planet Earth as a whole, these comparisons seem silly. But until (or if) we *all* think of our planet as one, international cooperation has to take into account these metrics to set targets and control changes. In other words, until we have a universal and unified set of regulations that make environmental externalities[†] part of the cost of doing business, we'll need to track by country. And when we track by country, there are important distinctions to be made.

We currently live on a planet with eight billion people, where roughly one billion live in extreme poverty (less than $2 US per day) and three billion live with an income between $2 and $8 per day (It's unlikely that you are part of this group of four billion people). Don't forget that living in extreme poverty means that your basic human needs (food, water, shelter) are not being met! If you live in the second group ($2–$8 per day), you are constantly fighting to not fall back into extreme poverty. When talking about things like environmental regulations, it would be unjust to stop economic development for these people, so we cannot therefore expect anything like universal restrictions that apply to every population equally. This is why in so many UN meetings and deliberations, they talk about reducing the carbon footprint in rich countries, while allowing lenience to struggling nations.

Remember earlier when we talked about how on average we would need 1.6 Earths to meet our current pattern of consumption and disposal? Well, the only reason this number is not as high as three,

---

* The Human Development Index (HDI) measures a country's development based on life expectancy, education, and income levels, offering a more holistic view of progress than income alone.

† Externalities are the negative impacts that do not have to be paid by the polluter, but rather shared by whoever has to deal with the burden later (a population not having clean water, for instance, or a government decontaminating a piece of land). We will explore this further in later chapters.

four, or five is because of these four billion people that have such small environmental impacts. To allow these people to develop beyond their current means, we, the $8-plus-per-day group, need to start working to lower our current environmental impact. We can expect to see growing levels of pollution arising from the development of worse-off nations— and, in many ways, this is a good thing, because it means the quality of life is improving for the millions of people living in extreme poverty.* But that means we need to make extra effort to contain the footprint of affluent nations to compensate for it.

THE PSYCHOLOGY OF CONSUMPTION

Wait, but can't we just wait for new technologies that will allow us to maintain our current lifestyles without harming the planet? Debatable. We have to try to lower our impact no matter what, and we really can't wait for a silver bullet of technology to come along and save us. First, because the timeline is uncertain—maybe such technologies will be ready in a decade, maybe in a century, maybe when it is way too late. Waiting is a bit suicidal. Second, it's not like we have to give up things that are indispensable—they are mostly luxuries, but ones that are harmful to the planet. It's a bit like smoking, in a way. You don't need it, and it is known to cause harm. So why do some people still do it? Because it feels good in the short term? Psychology is complicated...

----

* Obviously in an ideal world, worse-off nations would leapfrog into green infrastructure and green technology and have a small environmental impact as they develop. But for that to happen, there needs to be a global safety net, and this gets very political. But even in such an ideal world, the construction of houses uses materials that demand natural resources that have an environmental impact associated with them (think back to the example of concrete discussed in earlier chapters). We haven't found a magic solution to decouple economic development and environmental impact. We have ways to minimize the impacts at best.

One of the reasons giving up on certain aspects of our lifestyle is hard, in my view, has to do with the particular human psychology of relative comparison. Here's an example: My five-year-old smartphone works really well. I'm not exaggerating. There hasn't been a single time I couldn't do something I wanted because my phone was too old. However, everyone around me seems to have newer phones. Actually, most people around me upgraded about five times during the last five years. When I compare myself to them, I feel the urge to upgrade too. There is always that newer model with all the new features, the better battery, the new design, the new function that no one really uses but everybody wants to have. When I evaluate my phone in objective terms, it's fine. But in relative terms, it's old and obsolete. A similar case can be made for my car. Another for my clothes. Granted, I have some ten-plus-year-old T-shirts that my wife would love for me to refrain from wearing, but in absolute terms, they still do the job as well as my two-year-old T-shirts.*[153]

Yet another case can be made for salary. Australia has the highest minimum wage in the world. This can vary slightly depending on your method of calculation, but it will always rank very high. Scott Pape (an Australian financial counselor and author) did some quick math and found that the average Australian is richer than 99.7% of the global population![154] Yet, more often than not, one can hear in Australian conversation circles how they need to make more money, or how their salary is not "enough." Not enough for what? Who are they comparing themselves to? Based on the average Australian salary, only 0.3% of the planet's population makes more than them.

---

* Speaking of shirts, since the year 2000, the fast fashion industry has seen significant growth, driven by cheap manufacturing and the rapid turnover of clothing styles. This growth has intensified the environmental impacts, particularly in the countries where garments are produced. Key environmental issues include extensive water consumption, chemical pollution, CO_2 emissions, and textile waste. A transition back to slower fashion consumption is urgently needed.

Yes, the cost of living is high in Australia and whatnot, but my argument stands. I think the case against relative comparison is solid, from the saying "the grass is always greener" to academic studies evaluating how social comparisons in social media negatively affect one's mood.[155,156] No matter how pretty you are, you'll manage to find someone you think is prettier and compare yourself to that person. So where am I going with all this? Just bear with me for a minute.

Would you give up driving your car to work every day and take public transportation three out of five days of the week? You'd probably argue that driving is more convenient, the public transportation where you live is terrible, it would take too long, etc. You'd find lots of excuses, many of them valid, in order to keep driving your car. In a culture where driving is the norm, it's just the thing you do, like everybody else.

But what if that wasn't what everybody else was doing? What if the power of comparison and conforming could be used for good?

In the Netherlands people often bike everywhere—to school, to work, to parks, you name it. Everyone does it, on all parts of the wealth spectrum. Because everyone does it, the governing bodies are encouraged to make the cycleways better. And because the infrastructure is better, biking is actually more convenient and faster than taking a car (there are fewer car lanes because of the cycleways, less parking is available, etc.), so even more people use a bicycle, coming full circle to a positive feedback loop effect. The culture goes even further to support biking because buildings and cities are also planned to support this lifestyle (e.g., they don't have megacities where things are several miles apart), and people tend to live closer to work, support local businesses, do more leisure activities in their neighborhood, etc.

THE STEPS WE CAN TAKE

But wait. This book is not about sustainable cities (well, it is, in a way). We got here because we were talking about what you'd be willing to give up in order to be more sustainable, and how what you are willing to give up has a lot to do with what your neighbors have or don't have. There is an ongoing debate about what we as individuals need to sacrifice to achieve a lifestyle that Earth can comfortably handle. I have gone back and forth on my position about this. The fact is that the current lifestyle of a chunk of the global population is unsustainable. So it follows that it should be changed. But what exactly should be changed? Is it the lifestyle itself (in other words, the demand)? Or is changing the supply side more important? Should we focus on innovating the way we make things today so that they stop being a burden on the planet? The answer, I believe, is both. Let's begin by focusing on the demand side, and we'll return to the supply side of the equation later in the book.

While it could be argued that it's not realistic to expect huge changes from individuals nor to rely on such changes to make things right, and that expecting most humans to give up on lots of modern comforts is just unpractical, I would also argue that you are not most humans. You are reading a book about environmental sustainability, which tells me you are interested and engaged, and are surely capable of changing. When individuals make sustainable lifestyle choices the norm, it normalizes those choices for everyone else. As in the example above about bicycling in the Netherlands, these choices have a collective positive feedback loop, creating exponential growth on the level of social groups, families, communities, cities, states, countries, and eventually the whole globe. This momentum is what tips the scales for the needed social, economic, and political change that we'll talk about in the next chapter.

Going back to the question posed, yes, dealing with both the demand and supply side of things is important, but we need to keep in mind exactly how important each is, and where to most effectively focus our efforts. So, thinking about the demand side, what are some things we can, as individuals, change in our lives and sustain over time? Let's look at some little steps each of us can take right now.

Change transportation habits

When I first moved to Sydney, I took public transportation to my university. If there had been a direct route going straight from where I lived to my university, it wouldn't have been a long trip, but it was about 90 minutes each way. For me, the decision to get a car was a no brainer. Used cars are (somewhat) reliable in Australia and the fuel isn't expensive. I acquired a 21-year-old car that was falling apart but got me to uni in 45–50 minutes. That was 90 minutes saved on transport per day, which quickly added up to several days saved in a year. Eventually I moved closer to work and the drive was 25 minutes without traffic and an hour with traffic (talk about sitting idle in a car!). Public transport took about an hour.

I knew that several of my colleagues biked to work, but in my mind, they all lived fairly close—I didn't bother to ask them all, but the few that I knew rode bikes did so for about 5 km. One day, however, I was talking to a colleague who told me he biked to work. I knew where he lived, and it was 25 km away—with a lot of hills! I learned that he rode an electric bike, which made the long distance a little bit easier, but it still seemed like a monumental trek. Much like the story about my carnivore habits (coming up later in this chapter), something interesting occurred when I talked to people like my colleague who were doing it differently. Now there was a sense of possibility, a kind of "if he can do it, so can I" (maybe a little comparison/friendly competition as discussed earlier). His example convinced me I could at least try. My home was about 14 km away from work, roughly half of his journey. And so I tried. The first time, I got a bit lost and took some wrong turns here and there, but I made it and it was quite doable. I didn't have an e-bike, just my normal push bike, so I suffered a bit on the slopes, but I started going once per week. Then twice per week. And just like that, it started to become a habit. And then I eventually got myself a secondhand e-bike.

Looking back, this decision should have been a no-brainer too, just like my old car* was a few years before. It introduced exercise to my commute. It made commutes less stressful and more enjoyable. My carbon footprint decreased and the travel became cheaper—no fuel to buy and no toll to pay. And funny enough, my trip back home was faster on the bike than it was driving, due to traffic. Thus the average commute time remained the same as driving the car, with lots of added benefits. I could still take a car when it was raining heavily or when I had an appointment far away from work after hours, so I did not have to give up on any convenience. There are plenty of these no-regret changes[†] that leave us and the environment better off. But we often need a little push to get started because of our inertia and life's status quo.

Which no-regret changes could you implement in your life right now? Have a think. It may be that you've never thought about this, and once you actually try, you'll find there were possibilities all along. And hey, if something doesn't work for you, you can always go back to your old way of doing things.

At the same time as telling about how I started riding a bicycle to work in Australia, I also want to tell you about my experience riding a bicycle in Brazil. First, in a country with a large income gap like Brazil, purchasing a proper bicycle is no mean feat—depending on

---

* Which, by the way, is no longer among us. It was written off a couple years after I first bought it.

† No-regret changes are mentioned by the climatologist Michael E. Mann as those that make us healthier, save us money, and reduce our carbon emissions.

your wage, it may require a significant part of it. The cost of maintaining a bicycle, while way less than the maintenance of any car, can still be a burden. As is the case with several other countries, Brazil is still in its infancy in developing a bicycle-friendly infrastructure. There aren't many cities with bike lanes and, in the cities where these can be found, they are normally localized and cover only a fraction of the metropolitan area. In addition to that, bicycle theft is not uncommon. So, much like a car, you should have a proper garage-like spot to leave a bicycle overnight if you wish to keep it safe. Riding at night brings its own hazards. Not only because of the obvious low visibility that comes with nightfall (and which is independent of where you are in the world), but also because you could be mugged or assaulted. I for one rode a bicycle often in Brazil to go to classes and do chores close to home, but avoided at all costs riding in the evening and leaving my bicycle exposed overnight. All these caveats are not to say that people should not use their bikes in a country like Brazil, but that perhaps making the change like I did in an urban part of Australia is not as simple for everyone. Often only the privileged can afford to make this sort of choice. My point circles back to the changes we need to pursue in order to fight environmental problems: Everyone needs to chip in, but not everyone can chip in the same amount or in the same way. Much like the bicycle context, there are countries that can't afford not to use fossil fuels and whose path away from them requires assistance from better-off nations. Understanding the context is paramount.

The last thing we'll do in this section is run some numbers to understand the impact of flying, which you'll often find portrayed as the worst offender of carbon impacts (spoiler alert: it isn't). Let's take my own scenario for the sake of calculation. I live in Sydney, Australia, but my whole family lives in Brazil, and I am very close to my family. Not only my mom, dad, grandma, and sister, but also my uncles, aunts, cousins, and the list goes on. So I try to visit once per year, which makes my air travel carbon footprint a relatively high one.

As discussed in chapter 4, flying is responsible for about 1.9%–2.4% of global greenhouse gas emissions.[157,158] Consider our annual total of 51 billion tons of CO_{2e}, and we can calculate that flying accounts for 0.97–1.2 billion tons. A round trip from my hometown to Sydney emits anywhere from 4.2–7.1 tons of CO_{2e} (that's single tons, not billions or millions of tons), depending on whether I fly east or west.* So my trip contributes 0.00000035%–0.00000073% to the total. That number's tiny, isn't it? But let's think about it in a different way. If we flip the number, between 136 and 285 million people would need to take this flight in a given year to reach the total greenhouse gas emissions. Now that is a problem! We are about eight billion people. What this back-of-napkin math has shown us is that only 2%–4% of the world population would need to take this flight. And that is precisely the problem with flying. It is not a big deal when compared to the whole, and it is not even a big deal when compared to the contribution of the transportation sector. Yet, flying is practiced by such a small part of the global population (3%–4% on average) that it is a big deal when considering an individual's contribution.

You may recall our income brackets from earlier in this chapter and our discussion about how four billion people live on only a few dollars per day. Now we're talking about the other side of the Dollar Street. While a good chunk of the global population has a small carbon footprint, the flyers represent a small chunk of the population while having a very high per capita footprint. So should we just eliminate all airplane travel to solve the climate change problem? No. If you've been paying attention, the solution is more nuanced than to fly or not to fly. Practicing the little steps is important, and in an ideal world, we would all limit our habits to a sustainable quota that would be roughly

---

* I used simple calculators here, no fancy LCAs. I did use a few different methodologies just to be safe, but these numbers are by no means precise. We just want a ballpark idea.

equal. Yes, we should avoid flying whenever possible. We should also avoid driving when possible. But we could all stop flying altogether and yet the problem would still exist. It wouldn't even make a big dent (1.9%–2.4%!). So, sure, start using a bicycle. Avoid driving when you can. Avoid flying, too. But keep an eye on the big steps, many of which are outlined in the next chapter. And watch out for propaganda that shames you and others who fly or drive as being the big villains in this story. This is a great way to deviate your attention from the real deal. All the little steps are futile without the big steps, and these crucial steps depend more on systemic changes than on your choice of whether to see your family for Christmas or not.

Save water

Saving water may sound cliché, but perhaps we could run some numbers to make things interesting. First, the basics: We use water for everything, from drinking to bathing. Generally, we think of drinking water as pure water (well, "pure" water), but don't forget water is essential to any drink—from beer to juice, soda to whiskey. We need water to grow crops and feed animals, so producing our food also requires us to have water (by the way, remember that all foodstuffs have some percentage of water in them). Water's importance goes far beyond homo sapiens' needs as the water cycle is responsible for a myriad of geological and biological activities, which are in equilibrium in nature, in spite of being ever transforming. I don't think I need to convince anyone of water's importance, so I'll stop here. But why do we need to decrease our consumption of it? You may already be aware that only about 3% of the world's water is freshwater, and currently, only 0.9% of that water is actually surface water (the rest is locked up in ice or in the ground).[159,160] And that tiny and valuable fraction is precisely the water that is being wasted, as if we had an infinite supply. I am sure that eventually we'll come up with ways to desalinize marine water in a sustainable and efficient way, but as things currently stand, we ought to be careful with what is available. In addition, unless you are fetching water from a well yourself and pumping it manually, there is a whole process required to make the water arrive at your tap: storage, treatment, piping, pumping, etc.—all of which require energy and resources. So saving water also assists in keeping your general environmental impact down (including your carbon footprint).

Wait! But doesn't water follow the water cycle and always return? Yes...and no. While we can count on the total amount of water on planet Earth staying stable over time, not all waters are equal. The moment a city extracts freshwater from the catchment area and dumps it in the ocean, the freshwater no longer exists as such. Of course, some water will return to the original basin through rainfall, but not necessarily 100%. There is a reposition rate* that can be inferred by taking the average precipitation for a given catchment area. If the rate of extraction is greater than that of reposition, the amount of water for that catchment area decreases over time.† In a way, it is analogous to the carbon cycle. It's not that no carbon can be released. As a matter of fact, carbon release is natural and always happening (e.g., from organic decay or from the atmospheric-ocean exchange). The problem is the rate at which carbon (or water) is released.

---

* The reposition rate is the percentage of water that naturally returns to the original area, like a river basin, after being used and released, mainly through rainfall. Example: If a city uses 100 liters of water from a river and, due to rainfall, only 75 liters flow back to the river basin, the reposition rate is 75%.

† The rate of reposition is not the only thing that matters. As previously stated, not all waters are equal. A city grabbing water and discharging it downstream untreated, even if it returns at the same rate of extraction, may be also affecting the water availability due to water quality. There is a whole (huge) field of study around this, and by no means is this book intended to cover these important differences. The idea is to introduce the topic and encourage interested readers to go after additional information.

Let's now run some numbers. Not everyone has equal access to water, but let's aim for a world in which we all do. A person should drink between 2.7–3.7 liters of water per day.[161] What about toilets? How much water is needed to flush? Older toilets can use up to 25 liters per flush! Newer toilets and the introduction of dual flush helped lower the number a lot; 3.6 liters per flush is a good average (which already accounts for the ratio between single and dual flushes). So, keeping it simple, assuming five flushes per day, and that half the toilets are old (13–26 liters), and half are new (3.6 liters), that would be 18–100 liters per person per day for toilet flushing. Bathing varies greatly, but we can grab a range between three to nine minutes of running water per bath or shower. In Sydney, Australia, a six-minute shower uses about 50 liters of water. So let's say between 25 and 75 liters per person per day. Note how the volume of water for showers is a whole order of magnitude greater than that of drinking water, and how toilets plus showers dwarf the drinking water. Summing up the numbers so far, we need anywhere from 45 to 179 liters of water per person per day (at least until we change over all the old toilets and everyone starts taking more efficient showers), not to mention other uses like washing dishes and doing laundry. The current population is roughly eight billion people, so that's 0.36–1.43 trillion liters of water. If we reach the 11 billion population projected by the end of the twenty-first century, then the figures goes up to 0.49–2.0 trillion liters of water per day, or 180–718 trillion liters per year.

So what is the message here? Save water, right? Shorter showers? Flush less? Not really. Of course, doing any of these won't hurt, but in reality, our individual direct consumption is just the very tip of the iceberg (excuse the pun). First, because a lot of water gets lost along the way to its final destination. Leaks are a major source of water wastage. It is estimated that, yearly, 3.8 trillion liters of water are lost due to leaks in the USA alone. For the sake of grasping the magnitude of this, let's normalize* the population of the USA with respect to the world population (the 11 billion people we expect by the end of the century), which equates to 127 trillion liters per year. That is anywhere from 18%–70% of the water we need for human consumption! We can fix those leaky faucets in our homes as much as we want, but individuals certainly can't dig up the asphalt on their streets to replace their city's old pipes, nor can they eliminate the massive water wastage in factories and various industries—this needs to be done by governments and corporations.

But here is the real punch line: About 70% of the world's freshwater reserves are used in agriculture (mostly for irrigation)![162,163] This is the most important point I want to make here: The water consumption of individuals is not the bulk of the global water consumption. That is, if we all suddenly stopped showering, washing our hands, cooking and, you know, drinking water, we would stop less than about 30% of global water consumption. Doesn't that blow your mind? Think about it. All the water we "see" every day for everyday use is less than 30% of the total consumption. And that is an average! If you take my home country, Brazil, it is way less than that: Just about 10% of the total water consumption is for residential purposes (urban or rural). The vast majority of water is used in irrigation (almost 70%!) followed by animal use and industry— which together account for 88% of the consumption.[164] You may be aware that Brazil holds the largest freshwater reserves in the world. The country is bountiful in its resources, but not limitless. The last two decades have seen an 80% increase in the demand for water in

---

* For those of you who haven't taken a math class in a while, normalizing means I am taking the 3.8 trillion liters of water from the US and using it for the population of the whole world (that 11 billion projected population)—as if it were "normal." Of course, we know that the US consumes much more water per capita than a lot of the world's population, but regardless, global water wastage is very high.

the country, and additional demand is expected until the end of this decade.[165] The conclusion here is that individual action is important; it is, after all, part of the little steps. But keep in mind exactly that: It is part of the *little* steps. Much like transportation, as discussed earlier, even if we save all the water we can possibly save, we still won't make a big dent in total water use. The big steps involve tackling the bulk of the water consumption on a systemic level.

Save energy

The second law of thermodynamics tells us that not all energy is created equal. It adds to energy the characteristic of having quality in addition to quantity. And while that can often be neglected as unimportant at first glance, the truth is the exact opposite: While a joule of energy equals a joule of energy in the perspective of quantity, a joule of electricity is more valuable than a joule of heat—so electricity ranks higher on the quality scale than heat. This is because electricity can be easily transformed into other types of energy, while for heat, it isn't as easy. This can help you understand why your electric heater at home can convert almost all its energy into heat (efficiencies are close to 100% for almost all designs), while converting heat into electricity (using heat from burning coal, for instance) has much lower efficiencies, in the range of 30%–40%. Why does this matter? Because electricity has become the bedrock of modern society, and making it requires huge amounts of resources, infrastructure, and, well, energy. And as we saw in chapter 4, making electricity is the second biggest GHG emission group, right behind making things. That is why there is such a big push today to have electricity from renewable sources such as wind and solar, and to also electrify as much as we can. The latter allows us to make things using electricity (as opposed to fossil fuels, for instance), while the former means little to no carbon is being emitted to generate the electricity in the first place. But just because we are trying to electrify several carbon-intensive processes does not mean we have the luxury of wasting energy.

How to save energy? Well, you know. Try to turn off your stuff when you are not using it, like appliances, lights, etc. It is pretty straightforward. Do you think this is silly? Well, consider this: Roughly half of employed adults in the USA, UK, and Germany don't typically shut down their computers at the end of the workday.[166] One study claims that about a fifth of the workforce does not power down* their computers at the end of the day or during holidays.†[167] It takes seconds to turn off or power down your computer. How does that translate into global warming potential? I'm glad you asked. Based on the figure that 10,000 personal computers in the USA, UK, and Germany idly generate 1,870, 887, and 828 tons of CO_2 per year, respectively,[168] we can run some numbers. There are about two billion computers in the world[169], and about 56% of the adults employed in the USA use a computer at work‡[170] Let us once more take the USA as a proxy for the world (it is not) and say that 56% of the 3.4 billion workers worldwide use a computer.[171] Note that this math is a bit liberal, as we'd expect the global ratio of computers to be way smaller given that the USA is a high-income industrialized country. (You'll see where I'm going with this "radical" estimate soon.) Okay, so taking the average between the USA, UK, and Germany, we get roughly 1.2 tons of CO_2 per year for the idle time of ten computers. And now, considering the 56% of

---

* Powering down is different than shutting down. Powering down can refer to power scheme settings, sleep mode, or hibernation mode. Shutting down is literally turning off the computer so that no energy is being consumed.

† The source for this claim is not perfect, so take it with a grain of salt. It won't change the overarching conclusion if the correct ratio is between 10 and 30%, for instance. Same goes for other figures that I'll use in this back-of-napkin calculation.

‡ This is 20-year-old data, so definitely conservative considering how much every work sector has shifted toward online in the last two decades.

the hypothetical three billion that use a computer for work, we arrive at a figure of 202 million tons of CO_2. Big number, right? Let's now compare it to the real deal, that is, the 51 billion tons of GHGs we emit globally per year: 202 million of 51 billion is about 0.4%. Not that much, huh? And mind you, as stated earlier, this figure is probably an overestimation. We are assuming an upper bound regarding emissions, so much so that we are considering that about three-quarters of the two billion computers in the world are used by the workforce. What is my point? Powering down or shutting off our computers at work will not turn the tables on climate change. It will barely make an impact. But at the same time, how hard is it for us to do it? If using keyboard shortcuts, it will take you literally less than five seconds to put your computer in sleep mode. It's part of the little steps that add up.

But if you want to go a bit further than only turning off appliances when they are not in use, or powering down your computer when leaving the office or home office (as if there is such a thing as stopping work when you work from home), there are other easy ways to save energy. Changing your lamps from incandescent or fluorescent bulbs to LEDs is an easy way to save energy. LEDs are at least twice as efficient as fluorescents and at least seven times more efficient than incandescent! You noticed that I said *at least,* right? In Australia, there are even public programs for switching in which the cost of the new lamps are subsidized, and a nice person comes to your house to do the switch. Okay, what else? If you are into gadgets, there are smart meters that can be installed that assist by telling you how much electricity is being consumed, which equipment is consuming the most, and some models can even optimize the powering of devices according to the household's routine and the supply of the region's power system (by charging things when less demand exists, for instance). Don't go crazy installing smart meters, though. Too many of them can offset the savings they generate because of the impact of manufacturing them in the first place.[172] As a rule of thumb, install them for your main appliances, but maybe avoid installing them for your phone charger and hair dryer.

We have, as a society, come a long way in regard to energy saving. Just to give you an idea, things like sleep mode and energy efficient ratings only started in the 1990s, so they are quite recent initiatives. Of course, we are also using more energy than ever before. That is a fact. But I like to think we are becoming more aware of how precious energy is and how much of a waste it is to, well, waste it. Not only that, but we are more aware that the consequence of wasting energy goes beyond losing money, because it includes the environmental impacts related to the production, distribution, and consumption of electricity. Society went through a lot of research and development to be able to deliver electricity to about 90% of the world population[173]—let's not waste it.

Adjust eating habits

As I've already mentioned, I am Brazilian. Not only that, I'm from the southern part of Brazil where barbecue ("churrasco") is an important part of the local culture. I grew up going to my great-grandparents' house every single Sunday for a churrasco gathering with my family. We eat meat in basically every meal, so much so that not having meat in a meal is sufficient to not even consider it a meal. The meat dictates the dish. Everything else is a side dish meant to go with whichever meat you happen to be eating. I fit perfectly in this lifestyle, as did my family members and my peers. When confronted with the fact that eating meat might be one of my biggest individual environmental impacts, for a long time, I simply dismissed the idea of changing (denial, they call it). It was part of my culture, part of who I was. Changing would just be too hard.

When I moved to Sydney, I met a fellow Brazilian who would later become a very good friend of mine. Like me, he too was from the south of Brazil, and he too shared my customs and meat-rich history—except he was now a vegan. I'd met several vegetarians and vegans before, and have always been fascinated by the idea. But always as something for them, not for me. When I first met my friend and heard his story, it was interesting to see someone that came from the same context as me regarding meat diets. He never tried to "convert" me. As our friendship developed, all he did was answer my (often stupid) questions about his lifestyle and eating habits. I remember once asking him what he ate for breakfast: "Not just today, but every day. Like, what are your options?" Mine revolved around ham, bacon, sausages, etc.

I don't want to create any dramatic expectations here: I did not become a vegan. But I did change. Drastically. I went from eating meat in every single meal to not eating any meat for a while. I really missed it, though. It was a constant battle of self-control. After a while, I went back to eating meat, but instead of eating it 21 times per week (three meals times seven days—yes, literally every meal), I went down to five times per week, on average. It oscillates sometimes, but I never had meat for breakfast again, and most times that I'm given the option, I choose a vegetarian dish.

What is more, when faced with a choice of meats, I'll generally choose the one with the smallest environmental footprint. I keep saying "generally" because I've stopped shaming myself around this. Sometimes I really crave a steak or a burger, and that's okay. The point is I don't have one every day, or every week, or even every month. I've found a lifestyle that works for me, and it does not take effort to maintain. Moreover, when invited over for dinner or a party, I can eat whatever is being served. I keep the social side balanced with the "eco" side. And it always ends up being part of conversation because people want to know "what happened to me."

But now the real question: Does reducing or eliminating meat consumption matter? If so, how much does it matter in the grand scheme of things? You may recall from chapter 4 that food production is responsible for anywhere between 20 and 25% of GHG emissions, depending on how you group it. So it is a big deal. But the devil is in the details, and the details are worth exploring a bit more. The LCA tool we discovered in chapter 6 helps to answer the "how much does it matter?" question. The first thing to note is that the answer, again, is "it depends" (yeah, sorry). Depending on the metric you choose, the difference will be greater or smaller. For instance, the total impact of the food industry (not just meat!) in 2010 with respect to GHG emissions was about 26%, while with respect to terrestrial acidification it was 32%, and for freshwater and marine eutrophication it was almost 80%![174] So we need to be mindful of which impact is being considered. As a matter of fact, it has been found that there is a low correlation between different environmental impacts. In practice, this means that if we know that one food type is horrible in respect to one impact (e.g., farmed fish have a high eutrophication impact), we can't infer that other impacts will be similar (e.g., farmed fish have a high global warming impact). There is also great variability in how we produce food. So much so that the exact same food can have a very big range of possible impacts depending on how it is being produced. Among different food types, such variability is heightened. For instance, for the same amount of protein delivered, beef emits 42 to 125 times more GHG than peas. This figure considers the mean values, and there is a huge variability in the way these products are made. But even looking at the most efficient producers (of both products), the emission of peas is still 30 to 66 times less than that of beef.[175]

Yet another metric is land use. Agricultural land occupies 38% of the global land surface, of which one-third is used for crop land and two-thirds is used for grazing livestock.[176] So, if we talk about land use, the meat industry has a giant impact. Beef requires about fiftyfold the land required for crops such as peas, tofu, and rice on a per weight basis.[177,178] On the other hand, much of the land used to raise animals is marginal land.* And over 90% of what is used to feed ruminants (such as cattle) is nonedible for humans (e.g., husks)—about 37% of what we grow is nonedible (e.g., husks and straw).[179] I should point out that this last sentence is based on a 20-year old study from a researcher from the animal science department, which is not to discredit it, but rather to contextualize the data. The main point here is that this whole meat debate is not an easy thing to quantify and compare. The comparison needs to happen on a case-by-case basis, and these cases are often very specific to allow for a fair comparison.

Even the water-use estimates are hard to pinpoint. One can consider all the water that is used to raise the cattle, or only the water that has not returned to the water cycle (through the animal's urine, etc.), and everything in between. And just to make matters worse, determining the origin of the land being used can also be a game changer. Brazil is known for having land deforested to make space for beef production (or crop production). You can imagine how much worse the impact of producing beef there would be compared to, for instance, using already existing grassland for grazing. But how should this "much worse" impact be accounted for? Should we take all the animals being raised on such land and split the impact among them? And for how many years? Should we assume that the land will be used indefinitely for such purpose or only for a limited time? What about land that was deforested decades ago in developed countries? Is it fair that the impact of the meat produced there is lower than on the land where native forests were stripped more recently? Tough calls.

Finally, giving a single figure alone may be biased because different food types have different nutritional values (i.e., one kg of fish will

---

* Marginal land is land that has little or no agricultural or industrial value. This can be because it has poor soil or other undesirable characteristics. It is an economic definition because it is defined as land whose output would be worth less than the cost of renting/using it.

provide a completely different set of nutrients compared to one kg of bananas). And while this may sound obvious, note how complicated comparing food types becomes. What will we use to compare different foods? Calories? Weight? Protein? Some other macronutrients? A micronutrient? Depending on what you choose, the quantitative answer will vary. Thankfully, for the sake of our objectives in this book, the qualitative answer will mostly always be the same: Meat and dairy products have a higher environmental impact than nonmeat food types. And as a rule of thumb, we can assume that the worst is beef followed by meats such as pork and poultry, and that greenhouse gas emissions of plant-based products are 10–50 times lower than those of animal-based products.

I've tried my best here to balance the arguments in the meat discussion because it is a complex topic that is often dismissed as simple by many. Moreover, similar to flying and driving less, eating less meat is one of the things you can do personally to lower your own environmental impact. One of the most effective ones, actually. Food production as a whole has a significant environmental impact. Within the food types, the ones coming from animal sources *generally* have higher impact, not as high as some exaggerated claims you'll find out there, but high especially considering an individual's footprint. Among the things we can do to lessen our individual footprint, eating less meat or quitting meat altogether is one of the most impactful. However, and I keep repeating myself on this point, quitting meat alone will not solve climate change. Moreover, the idea of substituting all land used for raising livestock into land for growing crops is not realistic because of the land characteristic itself (e.g., poor soil) and because of the interactions that exist between the livestock and the rest of the environment (e.g., manure as fertilizer or eating parts of the crops that are nonedible for humans). At the same time, we have been witnessing the conversion of forests and other natural habitats into land for raising livestock to meet the globe's increasing demand for meat. This demand comes mainly from the economic growth of the "developing countries." The increase in meat consumption is actually a hallmark of such economic growth. And it is unsustainable for everyone to eat as much meat as is currently consumed in high-income countries. So, assuming we want people globally to have access to meat and other animal products, those of us in more privileged nations should start a transition toward consuming less of it.

Put simply: Eat less meat. Don't torture yourself in doing so, but reduce your meat consumption in a way that isn't too demanding on yourself (or else you may end up giving up right at the start) and improve from there. Start by adhering to popular ideas like "meat-free Mondays," or "meatless Mondays," which is a small step in the right direction. And then keep gradually reducing meat consumption to the extent possible. There is a whole culture around becoming a flexitarian, which is a semi-vegetarian diet and an excellent way to explain to people "what you are" (you'll get tired of doing this by the tenth dinner party). So just say you are a flexitarian as if everyone should know what it is, then walk away and let people google the meaning so they can pretend they knew all along. It works.

Speaking of eating habits, let me ask you a question: Do you prefer to buy and consume organic or conventional fruits and veggies (also known as industrial food)? Which do you think is best for the environment? As per usual, the answer is not straightforward. Generally, people associate organic with food that is healthier and more environmentally friendly. And to some extent, that is correct. But in our binary minds, we automatically assume food that is not organic must then not be as healthy and more damaging to the environment. This is somewhat justified given organic fruits and veggies are precisely those whose farming occurs without the use of synthetic fertilizers and pesticides, and without the use of genetically modified organisms (all popularly considered bad things). The catch in this situation is that the methods employed in growing/planting/making conventional food are much more efficient than organics, precisely because of the "bad" things aforementioned. This means that if we were all to start eating exclusively organic food from now on, lots of environmental impacts would be greater for many categories. For instance, regarding land use, eutrophication potential, and acidification potential, the impacts of organic may be double that of conventional foodstuffs.[*][180,181] At the same time, if we all completely relied on mass-produced food, we might end up having only a handful of places worldwide that harvest produce A or B, which is then exported globally. The impacts of transportation in these cases would be greater than if there were more hubs across the globe. Not to mention the threat to food sovereignty in several regions or the dangers of monoculture.

Hopefully, I've convinced you that neither of the extreme cases (either 100% of food being industrially produced or 100% organic) are optimal. In an ideal world, we would all be able to gather different food products from high-efficiency producers that are located as close to home as possible. That's not happening any time soon, but we can minimize our impact by combining the positive qualities of both organic and conventional foodstuffs, though some recent research shows the impact of transportation of food is tiny in comparison to making it. This discussion, while helpful in demystifying the "definite organic superiority" in terms of environmental impact, is not all that helpful in terms of the little steps we've been talking about. As we've seen, the potential impacts from these products is fairly similar, and such discussions often detract from more important differentiations that we'll address when we start talking about big steps.

---

* It's tough to generalize here, because the impact changes depending on the product in question (whether it's veggies, fruits, cereals, dairy, meat, etc.) and on the category of impact. For instance, the farming of organic fruit releases less greenhouse gases than industrial fruits, but it uses more land. In all other categories, the impact is not significantly different.

Consume less, consume better

I am fully aware this is not for everyone. Not everyone can afford the luxury of turning down cheaper products to buy quality ones.* But my guess is that you, reading this book, can. A good example is cars. If you buy a used car that is very cheap (let's say 25% cheaper than what you'd expect for the same model, make, year, etc.), it will most likely have some unpleasant surprises for you down the road. Maybe it will require more maintenance or changing parts more often, or it will consume more fuel. In the long run, the 25% you saved will slowly go down the drain with these added costs.

The fashion industry is perhaps an even better example, and one with huge environmental impacts. If you buy a low-quality garment, it may sound like a good deal at first, but after a couple of washes it will look worn out and you'll want to get a new one. Combine the price you paid for the first plus the second and you could have probably purchased a quality garment from the start, which will probably last you longer than the two combined. I'm not talking about super-expensive pieces of clothing that cost the price of a family sedan (true story, I'm not making this up), but about pieces that are made with quality materials, and are made to last. That might mean paying up to three to five times more for these items (I'm thinking of staples like T-shirts, jeans, sweaters, and shoes), but remember you won't have to replace them for a long time. I am probably an extreme example, but half of my wardrobe is a decade old.

I've recently had a problem with socks. I was running out and decided to buy some more. I thought I was getting good quality ones,

---

* There is a case to be made, however, that even if you are worse off, it makes more sense to buy quality products so you don't have to buy them again soon, rather than buy more affordable ones that will cost you more in the long run. Of course, there is a limit to this, which depends on how much you have to start with.

but one year later, the new ones were loose, and one had a hole in the toe. Yeah, I know, big problem. But my point here is that in addition to these new ones, I still have the ones I bought ten years ago (I kid you not), and they are still going strong. I disposed of the ones with the hole and will soon have to buy socks again, a mere one year after purchasing them. This is a small example of a repeating trend that applies to garments and to so many other things in life—from electronics to cookware to furniture, really almost anything you buy and consume. I can also tell you about a small speaker I bought cheaply that lasted me only two years, while my flatmate bought one from a quality brand at the same time and his is still working good as new, four years later. Not only did he save money in the long run, he also avoided creating hazardous waste that needs to be properly disposed of. I could go on with examples, but hopefully you (and I!) get the gist. Being aware of this can bring us, as individuals, to the little step of consuming less, which produces less waste. Please note, however, that waste is just the tip of the iceberg when it comes to low-quality products. The environmental impacts include the whole life cycle of the products: the extraction of raw materials to make it, manufacturing, transportation, packaging, and everything else in between.[*] The less individuals shop for low-quality products, theoretically, the fewer low-quality products will be made.

But changing is too hard!

Maybe you are happy with your lifestyle today and reluctant to give up any of your luxuries. I get that, I really do. But think about this:

---

[*] The argument goes further if we look at the social and economic aspects of our consumption. Fair trade products, for instance, are made to ensure all people involved in the manufacturing and supply chain of a given product are fairly paid and not working in precarious conditions. But this is a much more complex discussion that falls out of the scope of this book and my expertise.

By definition, you can't miss things that haven't been invented yet, right? But you can preemptively give them up. So if you never buy or consume *future* luxuries, you won't get used to them. Your consumption may not get better, but it also won't get worse.

Growing up, I was advised many times about the wage trap. You know the one: Your salary is 10, so you spend 10; your salary doubles, and you spend 20. Now you can't earn less than 20 to keep up with your lifestyle. All of a sudden, you are stuck doing something you don't want to do because you cannot afford to stop receiving the income you achieved. But if you are satisfied with your salary of 10, then it doesn't matter if your salary increases to 20 or 50. You can spend a bit more here and there, but your lifestyle is still 10. Likewise, if you are happy with your current smartphone, if it has all the functions you currently use, then you won't rush to change it up as soon as the new one is available. Do you really need all those new features? Does it really matter what kind of phone your neighbor has? Give it some time. Use it for longer. If it becomes faulty or too outdated, then switch. And maybe switch to a secondhand or refurbished one, or the third latest model, instead of a brand-new version of the latest model.

This slows down everything. If a large number of us changed our buying in this way, it would slow down production, manufacturing, and consumption (including resource consumption). You can apply this to most things in life. If you are happy with your current brand of jam, then don't buy the one that just arrived on the shelves of your supermarket that came from 20,000 km away (again, true story). Because once you do (assuming you like it more than your current brand), you grow used to it and become accustomed to an *unsustainable* practice. And the cycle of overconsumption continues.

I am not suggesting this approach is enough to turn the tides on humanity's harm to the environment. But if you are set in your ways and are not going to change your current habits (or think you can't), then this is a small thing you can do. You can keep the amazing life you have right now while consciously stopping yourself from falling into the "next level" pitfall, the "next level" lifestyle that is so easy to slide into and so hard to walk away from.

Keep watch

There are several cameras installed at the beach in Sydney, Australia, that constantly stream the ocean views online. Surfers take advantage of this by checking the condition of the waves before suiting up and taking the trip to the beach. It can also be used to check different beaches and choose the one that's the least crowded. The live stream allows you to get information in real time, and act accordingly. In a somewhat similar concept, some neighborhoods installed what has been dubbed a virtual community watch. Security cameras spread around the community are constantly streaming so residents can keep an eye out for any trouble or offenders, so they can monitor what is happening in real time and alert the police and each other to any abnormalities. In another similar, but twisted, endeavor, groups have installed cameras in some parts of the USA border to produce a live stream of routes that are commonly known to be taken by undocumented immigrants.[182] The free-of-charge stream is made available to the public, which means people from around the globe can spend their free time as border scouts and report any activity they see to the border patrol. And many do. Among the "active scouts," there are not only people in the USA, but also participants from other countries, such as Australia. Ignoring the question of ethics around such a setup, it is interesting to see people willing to give up their free time to "protect their country." What if there were a similar setup for environmental hotspots?

Well, there is. Environmental watch is a thing now, and I believe it has the potential to become something quite powerful. Monitoring the Amazon forest, for instance, has always been a huge challenge for the

Brazilian government (when the elected government cares, anyway). It's just too big of an area to allow for systematic monitoring. So much so that in many remote areas, one could illegally start a fire* that could burn unnoticed for a long time before being detected, and this detection would often happen through satellite pictures once the damage was big enough that it could be seen from, well, space. New technologies such as the use of drones have enabled authorities and local communities to spot activities like illegal logging in the region. But what if we applied the idea of the live stream cams so that monitoring of places like the Amazon forest was publicly accessible? Edge computing† advances could decrease the latency and bandwidth required to make this happen. I wonder if we'd have the same level of interest as people watching out for their neighborhoods or monitoring a state border. I sure hope so. Today, there is already a similar initiative using satellite images: The Global Forest Watch provides data on forests worldwide, which can be used by anyone (scientists, activists, governments, etc.) to back up claims and analyze the big picture. It's far from the live-on-the-spot detection system of the border or neighborhood watch, but it's a step in the right direction.

Let us suppose fires or illegal logging are indeed detected early and reported to the authorities—what happens next? This is where the individual contribution ends (little steps) and the collective begins (big steps). Any lucid person reading this book would expect the state authorities to act, right? Rush to the site, stop the fire, block the illegal activity, prosecute the violators. Yet that is often not the case. You see, while

---

* Fires are purposefully started to prepare pastures for cattle, but wildfires can also take place, especially with climate change increasing the forest's flammability.

† Edge computing puts storage and servers in the remote locations where data is being collected rather than requiring all of it to be transmitted, which is often impossible in remote areas of the Amazon.

we may correct the detection problem with technology and additional scouts (be it the general public or dedicated personnel), there is a deeper gap that needs to be addressed so our individual efforts actually amount to something. If there is no response team, not to mention sound policy and regulations, and if violators continue to walk away penalty free, early detection and intense monitoring won't change much.

Let me tell you the sad tale of my home country. As you may know, Brazil is the fifth largest country in the world in terms of territory. It is, like, really big. For context, it's larger than both Australia and the USA (if you exclude Alaska). Nothing against Alaska, but generally, people think of the USA as a big country and only think of the mainland. Brazil is also home to most of the Amazon forest, the remainder being split among neighboring South American countries. As you probably already know, the Amazon forest is kind of a big deal. Not only is it home to an astonishing number of species (i.e., has enormous biodiversity), it is also an important regulator of the climate, for both the region in which it is located and elsewhere. It takes up 60% of the Brazilian territory (at least from a legal perspective) and represents two-thirds of the tropical rainforests worldwide.[183] At the same time, Brazil's main economic activity is agriculture, mainly growing crops to be used to feed livestock, and exporting such crops and livestock worldwide—and it is quite a lucrative business. However, the more forest area, the less area for farming. So it is economically advantageous to decrease the forest area and increase the agricultural land (advantageous in the short term, of course). So what has been happening with Brazil's amazing forest? You know the story: the less forest in our way, the more money in our pocket. Who has the role of protecting this forest? Well, arguably all of us from around the globe. But we can simplify this discussion and just focus on the national authority responsible for the Amazon forest: the Brazilian federal government.

Looking back some 20 years, taking care of the forest by the federal government has never been a fairy tale. But between 2004 and 2013, the country was able to decrease illegal deforestation significantly by heavily relying on punitive measures.[184] In 2012, Brazil was able to reduce the rates of deforestation by an amazing 85% compared to the historical peak of 2004. The trend from 2013, unfortunately, has gone in the other direction, with 2019 being the highest of the decade.[185] In 2020, the deforestation rate was well above the target established by law in 2009.* Incidentally, the same year saw the deregulation or weakening of existing environmental legislation and a sharp drop in the number of fines. This led researchers to conclude that the 2020 federal administration took advantage of the pandemic to intensify a weakening of Brazilian environmental protection regulations.[186]

You'd think that after all the studies, data points, and media coverage (though the latter isn't nearly as large as I'd expect for the size of the problem), the Brazilian government would be acting and tackling the problem, right? Well, what if I told you that in June of 2021, the environment minister stepped down because he was linked to illegal logging in the Amazon forest and faced a criminal investigation? No, you did not misread what I just said—the minister, the top executive of the ministry of the environment, the very public department tasked with promoting sustainable development and protecting the environment was (allegedly) involved in destroying it. But the accusations were just the final blow; much damage had already been done. For instance, in May of 2020, the minister was caught on video in a cabinet meeting saying that the government should take advantage of the pandemic (a distraction) to push for environmental deregulation (he actually said to "rush the herd through," speaking metaphorically). As a matter of fact, a study on the topic evaluated the actions the Bolsonaro administration took between January 2019 and August 2020 and found 57

---

* And it is not just the Amazon biome. Other Brazilian biomes such as the Cerrado and the Pantanal have also seen deforestation associated with illegal fires and other related activities recently.

legislative acts aimed at weakening (!) the country's environmental protections, almost half of which were during the handful of months at the beginning of the pandemic when the population and the media were very distracted.[187] But at least this minister left, right? Things started to look better since, right? Right!? Don't get your hopes up. The next minister appointed had close ties with the big rural class in Brazil and used to be a member of the board of a lobby group for farming interests.[188]

The point of this story, as I've mentioned repeatedly, is that individuals can only have so much impact without the support of government leaders, policy, and other institutions. Let's pretend that, similarly to the previous examples, a 24-hour surveillance system was installed in the Amazon forest so that each and every one of us could scout the area constantly. What good would that do if no one answers the call to action when something goes amiss, if there isn't a proper infrastructure in place that will keep wrongdoers from practicing illegal activities for their personal benefit?

While patrolling the forests and reporting offenders may sound appealing to some, this idea of keeping watch goes much deeper. It is not necessarily to keep watch literally, but to keep watch metaphorically. It is our duty as individuals to be aware of the rates of deforestation, of species going extinct or being threatened with extinction, knowing which industries and companies are most responsible for polluting, and knowing which governments are causing the most harm through insufficient regulation or destructive incentives. Likewise, what are different corporations doing to decrease their impact and improve our planet's health? If they are doing something to revert the situation, is it enough? Or is it just enough to promote their brand as nice people who care, effectively greenwashing their way out? While our individual little steps may not be enough to turn the tides of environmental collapse, it is our awareness that leads to the action of holding those entities with powers accountable so they will make the big changes our planet so desperately needs. The little steps I've discussed in this chapter are the things we can change in our lives to help the environment, but these little steps alone will not solve the problem. This is contrary to what many of us have so often heard about the importance of lowering our individual environmental impact to save the planet. Many of us feel guilty for not doing enough, for not being perfect in our attempts to live sustainable lives. Like me, you may love a steak now and then. You may take a few flights a year to visit family, for work, or to go on vacation. You may take long showers or water your lawn or not always clean and recycle that gooey peanut butter jar (we've all been there). But I want to make the point that doing the little steps is not about doing it perfectly; in fact, striving for perfection is often the enemy of progress. When we feel like we're always failing (because, spoiler alert, it's impossible to be perfect, in both living green and everything else), it makes it more likely that we'll give up and stop trying altogether. Shame leads to apathy, and that's the last thing we want.

Rather, committing (as best we can) to taking the little steps I outlined in this chapter is, more than anything, about changing our mindsets. It is about living with our larger world as part of our consciousness, about beginning to understand ourselves as citizens of the Earth and acting accordingly. This change in consciousness sets an example for the people around us, influencing others to take similar steps. This is how change happens—from the bottom up. Our individual efforts alone may not be enough to save the environment, but they are necessary for changing our culture to one that will hold our leaders, governments, and corporations accountable. Our little steps are stepping stones for caring, awareness, and the big steps to come—which is the topic of the next chapter.

CHAPTER 8

BIG STEPS: CHANGING THE SYSTEM ITSELF

SOME FIFTEEN YEARS ago I flew to a nearby capital, roughly 500 km away from the town I lived in at the time. It was a very short trip whose only purpose was taking a test that was only available in this particular city. My flight arrived at night; I checked into a hotel, woke up, took the test, and flew back. On my way flying there, when I arrived at the airport, I decided to take the bus to the hotel. While waiting for the bus, a man (in his early twenties) who was also at the bus stop threw some rubbish on the ground (paper balls, as I recall). Everyone at the bus stop, including a friend of this man, saw this, and no one did or said anything. I wanted to, but had to build up my courage. At some point, the man got on his phone and, seeing my opportunity, I picked up the rubbish and challenged the friend by asking, "Did you drop this?" He denied it. I said, "I saw one of you tossing it away." He bluntly said, "I told you, I did not do it!" I threw the litter in the trash can (which was literally right beside the bus stop, probably five steps away). As soon as the first man got off the phone, he and his friend chatted about what happened. The first man looked me right in the eyes, grabbed more paper from his pockets, and threw it on the ground in front of me. A lot of things crossed my mind during those few seconds. Everyone at the deserted airport bus stop was watching the scene (the whole thing was very entertaining, and smartphones weren't ubiquitous then). I remember thinking (or did I say it out loud?) *This isn't even my city*, while giving him an evil grin and picking up the waste (again) to throw it away in the bin. My blood was boiling, but I kept calm. The bus arrived and life went on. I couldn't think about anything else for a while (guess how well I did on the test?) and, as you can probably tell from this recollection, the episode still lingers in my mind.

Why was I so furious that I decided to do something about it? It's quite obvious, right? We should throw rubbish in the correct place to avoid polluting our cities and to do our part. Throwing rubbish on the ground is an act we can directly relate to "harming the environment." I can only hope that, in addition to placing those pieces of paper in the right place, I also made the person think about what he did so he would not repeat it. He would not change his behavior for me, obviously, but perhaps it would make him think twice the next time. The same goes for all who were watching at that bus stop. Wishful thinking? Probably.

Let's take a step back, though. If I quantify the detrimental impact that the paper balls would have on the environment had I not picked them up and compare that to the environmental impact of the flight that all of us at that bus stop had taken, which is larger? No contest, obviously. The flight completely dwarfs the paper-ball waste. What about me? I was actually taking that flight twice in less than 24 hours! My negative impact was probably the highest there, dwarfing everyone else's by a good margin. My 15-years-ago self thought I was doing the right thing in that moment (and maybe I was), but my five-years-ago self has a much broader understanding of the big picture. Yes, I might have been doing the right thing in regard to those few pieces of trash, but that was negligible in comparison to the much bigger wrong thing I was doing: flying.

Today, my understanding is even more nuanced, different from both of those former selves.

Today I understand that the individual impacts of my flying are dwarfed by the impacts of corporations. Yes, as mentioned in chapter 7, flying is a big contributor when we think about impact per capita, but let us not forget that flying contributes to less than 3% of the total carbon emissions. Of course, one may choose to avoid flying as much as possible, but when we focus our attention so much on the little steps, the danger is that we will forget the big steps. In other words, if we focus too much on avoiding or reducing our 3%, we may forget about the remaining 97%.

This forgetting is precisely what a lot of corporations are counting on—and actively instigating through strategic deflection campaigns—the fossil fuel industry being the main offender. In the words of leading climate scientist Michael E. Mann:

> The fossil fuel disinformation machine wants to make it about the car you choose to drive, the food you choose to eat, and the lifestyle you choose to live rather than about the larger system and incentives. We need policies that will incentivize the needed shift away from fossil fuel burning toward a clean, green global economy. So-called leaders who resist the call for action must be removed from office.

Mann goes on to say that

> inordinate focus on individual action can erode support for systemic solutions to the climate-change problem—that is, governmental climate policy. Given that effective policy is far more critical than individual behavior in actually achieving the necessary carbon emissions reductions to stave off catastrophic climate change, the case could easily—I would even say convincingly—be made that attempts to redirect focus to

communicators' individual carbon footprints are antithetical to action on climate.[189]

As I've mentioned throughout this book, the big picture is precisely this: Individual action helps, but it is insufficient for significant change. Some, like Mann, even argue that it may be counterproductive if people focus their actions and activism on the impact of individuals, rather than on that of corporations and governments. To be sure, the ideal case is for us to do our part individually while also pushing for the required systematic change. This is the combination of the little steps and the big steps that will lead us to a sustainable future. But it is easy to get lost in the little steps (like the paper ball being littered at the bus stop) and forget about the big steps.

A powerful example of differentiating the big steps from the little steps is the matter of electrification. As I mentioned earlier in this book, one of the main pushes to combat climate change is to electrify our products and processes as much as we can. That is, convert processes that currently run on a fossil fuel energy source (e.g., making steel with a coal-fed furnace) to ones that use electricity. Same for products like cars, heating and cooling systems, etc.* That concept may seem abstract for some, but let's think about a modern kitchen, which has plenty of examples: In the past, water and food were always heated on a stove or in an oven, which was powered by some sort of carbon-based fuel. Today, modern kitchens of well-off families will often have electric appliances: stoves (by conduction or induction), ovens, kettles, sandwich presses, microwave ovens, toasters, air fryers, you name it. They all run on electricity. A lot has changed in the kitchen setup, and

---

* Processes and products here get a bit tangled. Cars can be thought of as transportation alongside buses, ships, trains, etc. And transportation could fit the process classification, which would put cars in the process box? Nevertheless, this distinction does not make any difference here. The point is that we are trying to electrify as much as we can, with both processes and products.

the idea is to do the same everywhere else, worldwide. But electrifying products and processes is only effective if the electricity comes from renewable sources, not more fossil fuels. Changing the energy matrix requires a big push against the status quo, and individuals don't have that power on their own, even if each and every one of us does the right thing. It requires organized movements, governmental change, and the embrace of corporations. Think about the powerplants we talked about in chapter 4. And who buys the bulk of cement and steel? Corporations and governments, not individuals.

My point here is that, while we can certainly do our individual part by switching over to electric products, the bulk of the change needs to come from institutions. And as you are probably well aware, individuals generally don't have a ton of power over corporations or governments. So the idea that an individual can have a great impact (even if we "all do our part") is simply not enough. The good news is that individuals do have great power in forming groups, and that kind of organized movement is what leads to government change and drives corporate accountability.

Please don't get me wrong. While I did separate the "Solutions" part of the book into two sections (little and big steps), they are very much interrelated. I do not want readers to be dismissive of the little steps just because they are, well, little. As I mentioned in the previous chapter, living with an environmental consciousness has a ripple effect that fuels bigger actions and inspires others to do the same. Living a life of integrity and making choices according to your values lifts you up and everyone around you. If we get stuck in apathy and feelings of powerlessness, we'll forget that we actually do have quite a lot of power collectively to influence what government and corporations do.

However, I also believe it is important to know how much each action actually contributes toward our end goal. After all, it would be pointless to mow a lawn with scissors only to find that when you got to the end of it, the first portion has grown back and is now twice the original height! If we get stuck only doing the little steps, thinking we are solving the problem, or if we think the problem would be solved if everyone just did the little steps, we'll be wasting our energy and time without ever reaching our objective. This is exactly what the worst offenders want us to do. The old divide-and-conquer technique is often the strategy employed by corporations whose interest goes in the opposite direction of environmental sustainability. For instance, get supporters of solar energy to fight wind energy enthusiasts instead of having them join forces to push renewables forward; get people who like driving to be anti-environment because they think activists are coming to take their trucks; get people who like eating meat to reject the idea of meatless Mondays because they hate vegans for shaming them. And so on. The big steps require organization, strategic thinking, and a clear understanding of the objective, which all come along with the little steps.

The little steps are important, and should not be forgotten or pushed aside. They are the actions that get you going, that start your momentum, that get you out of your comfort zone, and spark new possibilities in your mind. And the big steps may seem overwhelming if you haven't strengthened your capabilities and understanding before-hand. This chapter will focus on how we can turn the momentum of our little steps into the big steps that will create the significant change we need to save our planet.

SMART INVESTING

Do you have investments? Where? Better yet, what is your money doing? I get why you invested in the first place; like most people living in a capitalist society, you wish to either maintain or grow your wealth. But the question is not why you invested, but where you invested. And I'm not asking if you invested with bank Y or Z, nor am I asking if you invested in bonds or stocks. What I want to know is what your money is promoting while you lend it to someone else.

You see, we often forget that investing goes beyond depositing money and collecting interest. Whoever is borrowing money from your investment is borrowing for a reason. Perhaps it is a bank that is borrowing your money to lend it to people trying to finance a house (your savings account at the local credit union), perhaps it is for enterprises who are seeking to expand into a new venture (your stocks or mutual fund), perhaps it is for the government to use as it best sees fit (your government bond). But understanding that your money is going to serve a function other than to provide you with interest is quite an important step. I know it took me a while until this finally sunk in. Yet when you understand it, you can view your investments with another perspective. If you are someone simply looking to make the most money, you might not even want to consider what your investments are being used for. But given that you are reading this book, I would hope otherwise. I have a small portfolio with shares* and it took me a while to realize how controversial it was that I held shares of Petrobras (the Brazilian petroleum company). From a purely financial perspective, it was an investment that made sense, but going a bit deeper, it made no sense that I was part owner of a company whose main business revolves around extracting and producing fossil fuels. This is a very straightforward example and hopefully easy to follow, but sometimes what your money is actually being used for can be more elusive. A lot of times, people don't even know what specific stocks

---

* Shares (aka stocks) are parts of a company. When you own a stock, you partially own the company (i.e., you are a "shareholder"). Companies can choose to go public through an initial public offering (IPO), and from that point on its shares are exchanged in the financial market. This is what the NASDAQ, ASX, NYSE, TSE, Euronext, etc. are all about. You can find a lot of the companies you know listed in the market; e.g., Google, Facebook, Tesla, Apple, Amazon, etc.

they own if they're investing in a portfolio managed by a broker, money manager, or investment company. Same with savings accounts—it's a bit of a wild goose chase to find out what banks are investing in. All this to say it is worth paying attention.

And please don't get me wrong. I don't think that me selling a handful of shares will make a dent in the company or in the fossil fuel business. However, if there is a big enough movement pushing for divesting in fossil fuel companies, you can bet their stock price will decrease and that they'll listen. It may seem far-fetched, but the case of GameStop in 2021 shows it is not impossible. Even cooler is the fact that the GameStop case was mainly articulated by individuals who organized online, against hedge funds and large investors.[*]

If you skipped the last few paragraphs because you don't have any investments (or think you don't), I have news for you. "This is not my reality! I don't invest, I don't hold stocks, I don't have bonds," you may say, and while it may be true that you don't directly invest money yourself, it is more than likely that others do it for you or in the name of an institution that you support. If you have some sort of retirement fund, where does it invest the money for your retirement? If you work for an organization, where does it have its holdings? If you are hiring a service, perhaps an insurance company of some kind, where does it invest its capital? If you work or study at a university, where does it place its financial reserves, donation money, operating funds, endowments, etc.? If you have a bank account of any kind,

---

[*] In a nutshell (and poorly explained), the GameStop case led to the loss of billions of dollars by hedge funds that were hoping to short sell the stock, essentially profiting off a company's failure. The plan failed because of a buying and holding popular movement that started on the Reddit platform. The stock price soared, badly hurting any short-selling prospects. There is some controversy to the case as to whether it was indeed a popular movement or institutional investors who held and inflated the stock price, but I'll leave any further analysis in your hands.

where does that bank invest your savings? The truth is, unless you live completely off-grid, you are entangled in the financial system, like it or not. So why not make use of your voice (and money) to direct your investments into sustainable practices?

This may all sound overly complicated and overwhelming, but it might be simpler to make changes than you think. There are plenty of investment firms specializing in ethical or "green" portfolios. Rather than storing your savings in a for-profit corporate bank, you can choose a local, nonprofit credit union, which tend to invest back into the community rather than in the stock market. A lot of institutions already have divestment movements, and all you have to do is join them. If there isn't one yet, perhaps you can start it—turn your little steps into big steps through organizing. Harvard University, for instance, has its own divestment movement (it's called Divest Harvard; not creative, but it gets the message across). Do these movements render any results? In her book *This Changes Everything: Capitalism vs. the Climate*, Naomi Klein mentions several universities, colleges, and religious institutions that have announced their intention of divestment.[190] Speaking about the University of New South Wales, where I currently work, she says, "The University will divest from direct ownership and any commingled funds that include public equities and corporate bonds of companies whose primary business is the ownership and exploitation of fossil fuels reserves by 2025."[191] Note that now we're talking top dollars, not just the investment of an individual. But Klein says that the divestment strategy's main aim is not so much to hurt the fossil fuel industry financially as it is to erode its social license to operate, to make it morally unacceptable to partake in financing the industry, and to take away its political power.[192]

This is just one example of individual voices changing the culture of an institution—of little steps leading to big steps. As Klein said, these movements are often about eroding the *social license* of the worst offenders, of making it clear that their members (or customers, or investors, etc.) want them to behave differently. These kinds of pressures from stakeholders can add up over time, over entire industries.

So think about how this can apply to you. Where do you work? Where do you bank and invest? Where do you have a voice? Make use of it.

BECOME AN ACTIVIST: THE MANY FACES OF ENVIRONMENTAL ACTIVISM

People generally associate environmental activists with big grassroot movements and Greenpeace-sized NGOs. Or there are the stereotypes of hippies, radicals, greenies, and tree-huggers. While there is nothing wrong with any of these, the environmental movement is so much more than any of these things, and these generalizations are somewhat restrictive. We can recall some people already cited in this book who do not fit the aforementioned stereotypes, such as Al Gore in his corporate-like suits or Greta Thunberg who is a teenager and dresses like one. Activists for the environment, as we shall see, include entrepreneurs, lawyers, board members, farmers, investment groups, and the list goes on. Environmental organizations also vary greatly in the spectrum, from the ones that focus on individual change within the system to those that call for a complete reform of capitalism, with most falling somewhere in between.

My main goal in this section is to debunk the caricature of the environmental activist who distributes pamphlets, organizes protests, and is a fierce public speaker. Again, there's nothing wrong with that, and these kinds of activists are certainly needed, but you can also be an activist in so many other ways: by proposing enhancements in your neighborhood, using social media to shed light on environmental issues you find important, writing letters or calling your government representatives, and bringing up environmental topics in circles of conversation among your friends and family.

Environmental activists come in different shapes and flavors: You can be vegan or vegetarian, or not. You can be a car owner or not. You can live off-grid or not. What makes you an environmental activist is not checking all the boxes of what you think is best for the planet (though you'll probably check some), but fighting to change a system that is doomed to fail, in whatever way feels right to you—as long as you're doing something. It is important to understand that environmental activism goes beyond the short-sighted and often directly related things that first come to mind (you know: planting trees, saving a species from extinction, preserving a certain ecosystem); it's about fighting an entire system that simply does not prioritize the health of the environment and the species contained therein (human beings included!). It's about breaking the mentality of instant gratification and quick returns on investment at the detriment of future generations. It's about ensuring sound and just processes are taking place.

It's also about making sure waste is not being sent to countries that don't have the infrastructure to deal with it. It's about keeping people from contaminating themselves because it's the only way they can make a living. There is a notorious quote from a Brazilian environmentalist that summarizes this idea well: "Ecologia sem luta de classes é jardinagem" (freely translated, it means something like "ecology without class struggle is gardening"). The quote is from Chico Mendes, an activist, rubber tapper, and union leader who fought to preserve regions of the Amazon forest and advocated for human rights. His courage meant he was often sworn to death by many. Indeed, he was eventually assassinated.

As stated in *Our Common Future* (see chapter 2 for a refresher), there is a clear link between environmental conservation and social progress, and the former can only be achieved through a just and equitable distribution of resources within and among nations. Promoting social justice is a direct form of environmental activism. And to be sure, it does not matter whether you see yourself as left or right on the political spectrum, you can still promote social and environmental justice. Neither socialist nor capitalist ideologies hold the health of the environment as a priority, so debating political ideology is not the path to finding solutions. In fact, it is just one more way the divide-and-conquer technique has been used against us, so we'll keep fighting each other instead of the powerful groups responsible for harming the planet that is home to all of us. We need something different, an evolved system that is capable of accounting for the needs of the people and the environment at the same time. A sustainable system.

Moreover, movements like anti-apartheid, universal suffrage, civil rights, the (re)democratization of Latin America, the decolonization of the African continent, and India's independence (to cite a few) happen through the participation of many unknown heroes, by ordinary people like you whose names don't make it into the history books. I'm a big Noam Chomsky fan and this is straight from his thoughts.[193] The real movements, the ones that have the power to create unimaginable change, are happening right now in the background somewhere, at this very instant. You can be part of these movements by supporting or joining existing ones, or by starting your own. Each of these are huge steps that need to be taken if we wish to make a difference.

There are as many different ways to be an activist as there are people in the world. You can start by talking to your friends and relatives about significant changes and why they are important; by pamphleting (literally or metaphorically) about your cause; by boycotting brands and events, not just as a consumer but as an individual with a voice that echoes, or as part of a group; by organizing and thus multiplying the voices that call for immediate, measurable, and transparent change; by lobbying the government, be it local, regional, or national; by protesting as a citizen; by donating to worthy environmental causes— all of these efforts are part of our fight for survival as living beings.

Okay, that's beautiful and poetic, but how is it actually achieved? What can you and I effectively do? Let's get into the nuts and bolts of it with some examples to illustrate the big steps in action.

Boycott

Boycotting brands and institutions can take many forms. It can be done as consumers who stop purchasing certain brands that are not doing their part or that are dependent on maintenance of the status quo (e.g., fossil fuel, mining, or clothing companies known to have environmentally harmful production and unethical labor practices). It can be done as individuals who will not attend events in which the purpose is misaligned with theirs, or are associated with or sponsored by brands to be boycotted. It can be done as spectators who refuse to spend their time or attention with news outlets, websites, TV shows, movies, artists, sports, stars, etc., that are financed by or associated with brands to be boycotted. These actions would all count as little steps, and they are important, but the big steps are achieved when individuals are able to organize themselves and collectively become too large in number to be silenced or ignored.

An example of a successful recent boycott movement is the many social media-based endeavors that expose shady online corporate sponsorships and advertisements, the most famous of which were led by a group calling themselves Sleeping Giants. These boycotts peaked between 2016 and 2019, taking place in many countries and targeting such companies as Breitbart News, Coca-Cola, Ford, Banco do Brasil, Decathlon, and Leclerc. The group's goal was to fight hate speech and the fake news that goes against democracy and science, taking the "follow the money" approach to disentangle the hot mess of how these sponsorships work. The idea was quite simple, yet very effective. Anyone could collect evidence of a sponsorship related to a given outlet and report it to the boycott channels online. The person was encouraged to reach out publicly to the company, showing evidence of the sponsorship and tagging the boycott channel on social media. Often that was enough for brands to take action and disassociate themselves with the given outlet. The boycott channels often kept track of the progress and used their web presence to call out companies that failed to take action.*

Unfortunately, these kinds of grassroots boycott movements have lost momentum in recent years. While the tactics are most likely still effective, they require constant public attention and monitoring to work. And that is not how the internet works, unfortunately. We tend to move on quite quickly from one thing to the next. Moreover, placing the responsibility of monitoring each action of all the companies in the hands of individuals is bound to fail. Of course, as mentioned in the previous chapter, the public must keep watch to protect the common good, but the responsibility should mainly fall on strong institutions that actually have the power to effect policy and make sure regulations are being enforced. Where is these companies' money going? In which outlets are they advertising? How does the institution position itself—not how do they want the public to perceive their positioning, but how do they *actually* position themselves? Spend a few minutes searching online for your favorite brands, and try going beyond the paid ads on the first page

---

* For example, let's say you see an advertisement for company X show up in a website that displays fake news and clickbait titles. You spot the advertisement, take a screenshot of the page, and send it to the social media channel of company X saying, "Hey, I've just seen your ad on this website." Because of the way click-per-view works, company X could have been oblivious to it until this moment (or pretended to have been). So now that they have been made aware of it, they can contact their internet ad provider (e.g., Facebook or Google) and explicitly say they do not want their brand associated with the fake news website. Company X can be any institution that partakes in advertisement, from a business that deals with products to one that deals with services, from stores to universities to political campaigns. The boycott channels just assist with the visibility of the situation and with following up to see if indeed company X has stopped advertising on the dodgy website.

of your search results. Have there been articles from reputable sources written about them, either positive or negative? Have they been involved in any class action lawsuits? Have there been reports of sweatshop labor or environmentally harmful production practices? From there, you can make a more informed choice about what companies to support and which to steer clear of—and then take action by denouncing these practices publicly, alerting any applicable watchdog agencies, and calling on your online community to do the same.

Support green businesses

Speaking of companies and how they position themselves, have you ever heard of greenwashing? I started studying and learning about this before it had a name. Greenwashing, in a nutshell, is when organizations take advantage of the public's willingness to purchase "green" goods to do their part for the environment without necessarily doing anything meaningful—and sometimes doing things downright harmful! Examples may help here. These companies take advantage of our association with the color green, trees, and leaves with a general "good for the environment" vibe by making their packaging green and marketing with images of forests and trees to project the idea of an eco-friendly product, even if the product has exactly the same impact as similar products on the market. Another example is companies that advertise they are not using any CFCs (remember CFCs, the ones that damaged the ozone layer, that we discussed in chapter 6?) in their products, as if this is a huge ecological bonus they are doing because they care about the environment, when in reality using CFCs has been banned since the Montreal Protocol in the late 1980s.

Two independent investigations in Europe that analyzed online markets found similar results: They had reasons to believe that in a whopping 40% of cases, the "green" claims were exaggerated, false, or deceptive.[194,195] This is a situation where life cycle assessment (which we learned about in chapter 6) can surely assist. It's a great tool to understand the real impact of a product and to counter misinformation and greenwashing. But it still requires a massive effort in terms of institutional standardization of practices and units, and consumers having access to all the information about the supply chains of different companies. So yeah, when we talk about how companies *actually* position themselves, it is worth having eyes wide open and a hint of skepticism when evaluating their claims and advertisements.

The good news is that great ways exist to help consumers tell if a company's claims of being "green" are authentic and trustworthy. These often come in the form of various certifications from independent third-party organizations that do their own investigations into the products, services, and practices of a company to see if they should qualify as sustainable or otherwise Earth friendly. For instance, when it comes to building, a LEED (Leadership in Energy and Environmental Design) certification is the dominant trusted green building certification in the USA, while BREEAM (Building Research Establishment Environmental Assessment Method) is more popular in Europe. You may already be familiar with B Corp certification, which is an international certification that judges companies in the areas of governance, workers, community, environment, and customers to verify that they meet the highest standards of overall social and environmental performance, public transparency, and legal accountability. Other trustable certifications are Green Business Bureau, Fair Trade, Safer Choice, Green Seal, Energy Star, Rainforest Alliance, Forest Stewardship Council, and ISO 14001. I encourage you to do your own research on these certifications and others.

Protest

When you think about protesting, what comes to mind? Do you picture thousands of people marching? Possibly megaphones, mantras, shouting, and loud crowds? Conflict with the police or with opposition groups? Maybe violence? If so, let's quickly look at the Fridays for Future movement again. Recall from chapter 2 the movement that started with fifteen-year-old Greta Thunberg sitting outside parliament and striking in protest. Countless Fridays for Future events have taken place since that first protest in 2018, large and small, by young people all over the world who have been inspired by Thunberg's example. As evidenced by these activists, protest can be a simple, even quiet, act. Moreover, it can be nonviolent. There is data that points to nonviolent protests being more effective than violent ones; in other words, nonviolent resistance is more likely to achieve its objectives than its violent counterparts.[196]

For instance, we cannot count on individual action to lower the environmental impact of making the main materials responsible for current CO_2 emissions, because consumers in general do not buy cement or steel—large corporations do.[197] So this is something that needs to be tackled by organizations that are either pressured by groups of individuals or mandated by legislation. This is a good example of little versus big steps. As I've mentioned before, there is only so much an individual can do as a consumer when trying to send market signals to these big companies. Yet collectively, a group can pressure corporations or governments to commit to buying products that, while perhaps more expensive, have a lower environmental impact (e.g., lower carbon emissions, less toxicity). In the case of the government, an even better outcome can be reached by demanding policies that benefit the environment and hold corporations accountable, which we will cover later. Most governments and companies hate doing things deemed unpopular by their constituents and customers, so they are more likely to change when they see large groups of people protesting their policies or business practices. Protest then has the added positive effect of inspiring others to join who may have been on the fence before, but once they see the size and energy of the movement, they want to be a part of it.

Donate

My hope is that people will read this book and be inspired to actively participate in environmental movements, though I understand that it requires a lot of effort and time. You may want to be downtown for every protest or on the phone calling your local politicians, but the truth is you probably have a job, maybe kids, and other valid things that take up your time. While active participation is incredibly powerful and gratifying, and something I encourage you to do as often as you feasibly can, there are other ways to contribute if time is not on your side.

Active participation is important, but many of these small protest groups and environmental advocacy organizations also need donations to keep going. In this sense, there is nothing wrong with outsourcing the work to people who are willing and able to fight for a cause. But this is not an excuse to get complacent. Activism is not something you can charge to your credit card every month and forget about. It is not just about money—you are supporting an ideal. As the movement takes shape and gains traction, you want to be following closely so you can jump in actively when you feel the time is right.

But as we all know, money is power, and if you have it, I hope you will use that financial power to support the causes that matter to you. For instance, supporting established and respected environmental organizations amplifies your voice because governments and companies are more likely to listen to these large groups rather than individual people. Also, donations can go toward funding lawsuits against environmentally harmful corporations and even governments. Donating to preservation societies goes toward buying up land so that it will never be developed. The list goes on. Not all environmental organizations are created equal, however. Some of them have incredible overhead costs, so the bulk of your donation may be going to things like marketing rather than the actual activism you want to support. I encourage you to do your research and look for actual data for what these organizations have accomplished—not just their promises and ideals. Do some digging about how the money from donations is used. There are several independent online resources that rate and review various charities.

Educate

Put simply, the little steps are all about educating yourself; the big steps are all about educating others. In his book *The New Climate War*, climatologist Michael E. Mann argues that the institutions interested in maintaining the status quo and hindering the transition away from a carbon economy are now using different tactics versus those used in the past. These days, they're relying on what Mann calls the four D's: denial, doomism, delay, and deflection (as we've seen in previous chapters, environmental *deregulation* could also be included as part of the menu). Denial and delay are pretty self-explanatory: "There is no such thing as global warming," "It's not as bad as they're saying," or "We'll deal with it later,"* that kind of thing. The first sentence was prevalent in the earlier days of "the climate wars," but the latter two are relatively new tactics. Doomism means presenting the challenge at hand as something so big it creates a feeling of hopelessness and paralyzes action; i.e., it doesn't really matter what you say or do because the amount of change required is beyond the capabilities of an individual and there is no way we can get a handle on the situation. Finally, deflection revolves around the idea that the problem has little to do with, for instance, giant corporations extracting fossil fuels from the ground or the exploitative setup of the global economy, but instead, it has mostly to do with individual consumers filling up their gas tanks or failing to recycle that peanut butter jar (e.g., "Shame on you for eating

---

* Parents employ similar tactics with their kids using the "we'll buy it on our way back" delay approach so the kids forget about it. It works most of the time, I am told.

meat and not driving an electric car; this is all your fault!"). As if the responsibility is not in the hands of these huge corporations, but in the way people are living their lives. Mann's main argument is that today, after failing to convince the general public that the problem does not exist, corporations are doubling down on promoting inaction over everything else.

Here we circle back to the idea of having conversations with people around you about the issue. Discussing the problems, the consequences, and the paths of action are important (note this is the structure of this very book). There is no need to get technical about it, but rather point out in tangible ways how these issues affect their lives. I've found it is incredibly powerful to stress how the tactics of doomism and deflection are being purposefully employed to suppress engagement and action. I know I would have loved to have that pointed out to me before studying it myself. I have been working in the area of environmental engineering for a while now, and have often felt paralyzed by the numbers I found and the scale of the problem. In this sense, books such as Mann's were an eye-opener and helped me see these tactics for what they are: skewing truth as a means to deliver the lie that change is impossible. These tactics amplify the problem (e.g., climate change is a huge challenge), while leaving out the fact that it is solvable.*

As far as educating others goes, who should your audience be? As you have probably learned already, engaging in fierce discussion over

---

* As a personal example, I have always considered limiting the global temperature increase of 2°C as the ultimate goal. And of course, it is. But I thought of it as all or nothing. It's true that the warmer it gets, the worse it gets, but it's not like after crossing the 2-degree line, the Earth will implode. Things will get worse, feedback loops may go crazy, but this can also happen at a 1.8-degree increase, or at 2.5. Things will indeed worsen as temperatures increase, but all our efforts combating climate change are valuable regardless of the current increase shown by the "global thermometer."

social media comment sections is not very fruitful. No one's mind is getting changed there, and you may just be talking to a bot anyway. The audience you should aim for is people who are not hardcore climate deniers, as those people have already demonstrated that they are unwilling to listen to reason, evidence, or facts. They are also a much smaller portion of the population than you may think—they just happen to be incredibly loud clickbait generators, so they get far more attention than they deserve. The audience to focus on instead are people whose minds are still open, who may still think there is actually a debate going on about whether climate change is real, people who believe there is a problem but think it's either not too serious or is too far away in the future, and people who hear the news and think the battle is lost already. This is the vast majority of the population, and these are the people that can and should be educated, because once they understand what's going on, they may be inspired to take action, too. Talk to people who are honestly confused due to the various misinformation campaigns—even if they don't know they are. In a recent conversation, my friend correctly pointed out that it's hard to know what is true and what isn't because conflicting statements often show up in reputable outlets. This was a great opportunity to talk about how science works, how scientific papers that were peer-reviewed and published can contradict each other, and how everything we claim to know always comes with a level of uncertainty associated with it (recall the first chapters of this book). The idea is to find the consensus, which allows us to be more confident in our findings. Not 100% *sure,* but less *uncertain.*

If you happen to have the opportunity to formally teach, that is also of extreme value, be it through workshops, conferences, talks, or lectures. This doesn't necessarily mean in an academic setting. This can be done (and I would even argue it is more powerful) outside of a standard teaching setting. You can do a talk during a bring-your-parent-to-school day, through community-organized events, YouTube videos, podcasts, book clubs, family gatherings, you name it—wherever you feel comfortable and have the opportunity for teaching some of the basic principles. Depending on your audience, this can be more technical (you know, radiative forcing, greenhouse effect, the water cycle, the carbon cycle, the nitrogen cycle, etc.), or it can be more general. The point is to provoke thought, to inspire people to go after information on their own, become interested, or simply be more aware of why the environment matters and that they have a role to play in saving it (and that the role is probably not what they think it is— you know, less about buying an electrical vehicle and more about the big steps).

PRESSURE THE GOVERNMENT

Pressuring governments may sound like the most obvious thing. And it is. But we live in times when the obvious must be stated...clearly stated. It also does not hurt to revisit the ideas here and expand the scope of what pressuring governments entails.

The most obvious way to pressure the government is by choosing leaders who are willing to act on behalf of the environmental causes you care about. Let's just say that this speeds things up. But there is a lot of confusion and misinformation in the political arena, and finding sincere candidates is by no means an easy task. First, because there are the insincere politicians who say one thing and do something else. But beyond those, there are the sincere yet confused or misinformed ones. Disentangling the climate crisis and all the other environmental threats is a complicated task, and it often happens that leaders with good intentions end up inadvertently pushing for wrong policies.[*]

_____

[*] By wrong policies I mean anything from ineffective policies to policies that actually hinder the very cause they were trying to support.

Because of this, it will be useful to briefly go over some basic environmental policies so you can keep an eye out for them and use them as a sorting mechanism when considering candidates: Know when to listen and when to run away from them.

Decarbonize the electricity grid

One of the main things we need today to fight climate change is the complete decarbonization of the electricity grid. This can be achieved by harvesting as much as we can from renewable energy sources and offsetting the carbon from electricity produced from other sources (this leads to the famous term *net zero* you may have heard thrown around). Next up is electrifying as many processes, products, and services as we can. After all, what good is having a beautiful, renewable energy portfolio if most processes still run on fossil fuels? These two approaches, decarbonizing the grid and electrifying processes, are complimentary and will most likely happen in conjunction with each other,* but the overarching idea is to achieve net zero electricity first and then move things away from fossil fuels and into electricity.† An example most people think of is the move from internal combustion engines to electric vehicles, but there are plenty of large-scale industrial processes that have a much greater environmental impact, which many people are not aware of. This is a perfect example of the deflection tactics we discussed earlier—convincing the public

---

* As more things electrify, the bigger the incentive to deploy more energy farms. Policy also comes in both flavors and, again, these complement each other.

† If we magically moved everything to electricity today without first achieving net zero electricity sources, we'd pump way more carbon into the atmosphere than we currently are, because we'd need to use heaps of fossil fuels to create the electricity required by such demand. So, ideally, we should move to a clean grid first, and then switch whatever still relies on fossil fuels.

something is their fault, when really the most egregious offenders are the big corporations.

Finding politicians that embrace this change is a good start. But it would be political suicide* for them to embrace such an endeavor without also having some sort of social safety net or retraining program to assist the workers displaced by such a herculean switch. The Yellow Vest movement in France portrays this very well.† If a politician simply says they are in favor of changing the energy matrix without backing up the statement with a strategic plan that also includes the people affected, steer clear. Regardless of whether their words are empty or naïve, nothing will happen through the actions of such a person because they either won't follow through or their plan won't get the support it needs once they're elected. And this is true for both right-wing and left-wing politicians, just in case you are wondering.

Elect the right people. Protest and contact your politicians to let them know what you think. Make them change the energy matrix. Of course this is deceptively simple. How can the government possibly make this huge systemic change? It is not like they could simply tax carbon emissions, right? Well, I'm glad you asked. Meet the carbon tax.

----

* Political suicide means making a decision that will likely cause someone to not stay in power for long, either because their actions do not resonate well with voters or because they do not resonate well at the top with the political party or the wealthy corporations in power. So even if one knowingly and boldly takes actions on a sort of kamikaze mission, the changes—if they even occur—can be reversed shortly after by the newly elected person/party.

† The Yellow Vest movement began in 2018 as a protest against rising fuel taxes, which were introduced as part of the government's environmental policy to reduce carbon emissions. The protesters claimed that the taxes unfairly burdened low-income citizens, particularly those in rural areas who rely on cars. While the movement was rooted in economic grievances, it highlighted tensions between environmental policies and social justice.

Regulations

Many economists favor the carbon tax as the best method to accelerate the transition away from fossil fuels. It is straightforward and theoretically allows for a self-correction by the market. It is a way to operate within the existing system rather than requiring a complete overhaul of the global economy. In theory, it is wonderful. The main problem? It is very hard to implement politically. Promoting a carbon tax is, generally speaking, political suicide. And even the bold who pursue it for a while eventually back up a step or two along the way. Even when the carbon tax has been implemented as a cost-neutral policy,[*] it did not generally resonate well with voters.

Carbon tax is quite simple in its fundamental principle: It puts a price on carbon emissions.[†] Or we can also think it as putting a price on *externalities*, effectively ending them, by definition. Allow me to take a quick detour to talk about what externalities are before returning to the carbon tax.

Externalities (or external costs) are any consequence of a process that is not directly accounted for in the cost of the process and that affects other parties who did not choose to be involved in that activity.

---

[*] A carbon tax cost-neutral policy means the tax is designed so it does not increase the overall financial burden on taxpayers. The revenue generated from the tax is used to offset other taxes or is redistributed back to the public, often through rebates or reductions in other taxes. The idea is that while the carbon tax imposes a cost on carbon emissions, it is balanced by reducing costs elsewhere, so the net financial impact on individuals and the economy is neutral.

[†] Or, in slightly more complicated terms, a carbon tax is a fee that governments place on companies and individuals based on the amount of CO_{2eq} emissions they produce. Examples include taxing individuals directly for gasoline/petrol when they fill up their car, taxing both individuals and businesses for electricity from coal, or taxing companies for emissions that occur during the manufacturing and transportation of their products.

Economists call externalities outputs that are a by-product of a process that the creator does not take responsibility for. For instance, if a company produces leather, its internal costs include buying the raw materials and the necessary chemicals, paying the employees, renting the land, buying the infrastructure, and everything else involved in the process (electricity, gas, transport, packaging, etc.). In return for the input, it can sell a product. The inflows and outflows of such a business are easy to see. But the chemical processes used generate effluents that are hazardous. If the company can get away with dumping these chemicals into the regular sewage, it does not have to pay for the cost of properly treating/discharging the chemicals. However, there still is a need to treat the chemicals if we want to avoid polluting the environment and possibly causing harm to fauna, flora, and our community! In this case, the cost of treating such dumped chemicals is *external* to the costs of the company producing leather, because they are not paying for it. If left untreated, the cost will most likely increase by a couple of orders of magnitude since collecting all the (now dispersed) hazardous chemicals has a cost, as does decontaminating the land and water, and possibly treating people that became ill due to the inappropriate dumping. This cost is external to the company and, in this case, the cleanup costs would most likely be paid for by the local government responsible for the sewage system in which the chemicals have been dumped.

The leather company is a straightforward case. In most places, it is illegal to dump hazardous chemicals in the regular sewage system and companies that are caught doing that have to pay a big fine (mind you, that is *if* they are caught). But externalities sometimes come in more subtle ways. A good example of this is the myriad impacts of fossil fuels. At the time of writing, about 60% of the electricity mix worldwide is based on the burning of fossil fuels.[198] Not only does releasing carbon into the atmosphere create a massive externality that the global community has to pay for, but there are countless more direct examples. For instance, when the burning of coal releases soot into the surrounding environment, this directly impacts the lives of living beings around the power plants. They breathe air of lower quality, which has been correlated to respiratory problems, which in turn can be translated into medical costs (for the government or for the individual).[199] Here's something even simpler: Perhaps a store owner's store is constantly dirtied by the soot, so they have to allocate extra resources and time to clean it because of the power plant far away from their property. The power company does not have to pay for the store owner's medical costs or the cleaning of the store. While one may argue that the plant is not responsible for 100% of the cost, it does not pay even a fraction of the cost to cover these externalities. Moreover, note how complex calculating this fraction would be. The ramifications are often too complex for one to account for them all. And here we are talking of a single output (soot) of a single source (one power plant). Now imagine trying to map this for the whole economy of a country!

Okay, back from the detour. The carbon tax is a powerful way to create some balance with the externalities caused by fossil fuels. When speaking of generating electricity, for instance, the carbon tax effectively removes the financial incentive of using fossil fuels by increasing the cost of using them. However, raising the price doesn't in itself make the offenders pay for the harm they cause or have already caused, nor does it directly address the damage. And it only addresses one specific issue: carbon emissions.* As mentioned, it is also very hard to implement politically.[†200,201]

---

* Obviously this one specific action has several ramifications, but nonetheless, the carbon tax does little to address eutrophication, for instance, or the store owner's soot problem in my previous example, for that matter.

† Some countries have had success implementing carbon tax policies in recent years, with Finland, Sweden, Norway, and Switzerland leading the way. But the effect of these efforts on global emissions is negligible, and in the case of carbon taxes in South Africa, lower income people are disproportionately paying the price.

Despite the political hurdles, to date, 40 countries have implemented some sort of carbon tax. A few examples include Sweden in 1991, Canada (British Columbia) in 2008, and South Africa in 2019. Sweden is the poster child for the carbon tax because, since its implementation, the economy of the country grew by 60% while carbon emissions were reduced by 25%.[202] The country had the highest carbon price (north of $100 USD per metric ton of CO_2) for a long time, a title that has recently gone to Uruguay. There is a growing amount of empirical evidence suggesting that carbon taxes can effectively reduce carbon emissions or at least curb their growth without harming economic growth, employment rates, or competitiveness.[203] Nonetheless, implementing such taxation with justice and equity in mind is paramount, or else it can backfire like in the case of France and the already cited Yellow Vest movement, or in the case of South Africa, for which a 2021 study suggests the tax has disproportionately burdened lower-income people.[204] This makes the carbon tax even harder to implement—it's not just the creation of a new tax, which is by itself something that is not generally seen in a favorable light by the public, but it also needs to be designed and politically negotiated in a way that doesn't exacerbate existing social and economic inequalities. Bottom line, it is very hard to implement politically.[205,206] So other options are generally negotiated and agreed upon (if at all) as a form of compromise.

Cap and trade is an example of one of these compromises. Governments and regulating authorities cap, or limit, the total emissions across a given industry for a given year. Companies that find it easier to cut down on emissions can sell their "excess reduction" to other companies. This way, in theory, companies that can reduce their emissions for the lowest cost have an incentive to do so beyond their required targets, then sell such reductions in the market to companies who are not willing or whose reduction would cost more than paying in the market. Basically, the government states the reductions it wants to see and lets the market figure it out by itself. You can think of it as a stock market that trades carbon credits instead of company shares. I am personally not a big fan of cap-and-trade policies as they allow some companies to keep their single bottom line and the general status quo by simply throwing a pile of money one way or another. But these policies have been successful in several instances,* and I must admit they are definitely better than not having any policy at all. And generally the caps are reduced every year, so it gets harder for companies to achieve the target unless they have a long-term plan to lower their emissions over time.

In addition to carbon tax and cap and trade (which are quite universal in the sense that they target the market/country/region as a whole), specific regulations can also be implemented. Regulations can come in many shapes and forms, from ones that indicate exactly what needs to be done and how, to those that only set targets while allowing the market to choose its path to achieve such targets. The latter are more flexible and allow for the market to "do its thing." Examples of regulations include renewable portfolio standards, which, in a nutshell require that a specified minimum percentage of energy sold comes from renewable sources (and this minimum is increased periodically). Another example is regulations that limit the carbon that is released when producing, distributing, and using (burning) fuel.[207]

So far, we have talked about policies specifically designed to curtail carbon emissions, but regulations can be put in place to address all sorts of environmental issues. These include regulations that prescribe specific collection rates and recycling rates of waste (like the Waste Electrical and Electronic Equipment Directive in the European Union),

---

* The classic example here is that of Europe, with its 2005 European Union Emissions Trading System (EU ETS), which covers power plants and various industries and has lowered emissions by over 35%. Another good one is that of China's National ETS, which is the biggest carbon market in the world and a central part of the government's strategy for achieving net zero by 2060.

or limit the amount of pollutants in the outdoor air of a region (like the Clean Air Act in the USA). Regulations can also restrict certain substances from being used, or limit their usage. In the European Union (EU), since 2003, a directive that goes by the acronym of RoHS (Restriction of Hazardous Substances) sets the limits of substances deemed hazardous for different devices and products, such as lead. Interestingly, when the EU implements such regulations, this has a ripple effect in many other parts of the world. A computer manufacturer located in the USA does not have to abide by the RoHS directive—unless it wants to sell computers in Europe. And they all want to sell computers in Europe. So the regulation in Europe ends up reaching way beyond Europe itself. So much so that China, for instance, passed its own RoHS not long after its European counterpart went into effect.

A word on regulations: No one in any industry likes regulations. They make business more difficult, more expensive, and slower, and they inadvertently introduce the possibility of corruption as companies try to find creative ways to bypass the rules. Regulations tend to increase the cost of doing business, which often results in higher prices for the consumer. There are plenty of reasons to argue against them. Yet, after studying the topic deeply, I believe they are crucial to achieving a sustainable future. The reason has a lot to do precisely with the externalities we have just discussed. Profit-seeking companies will often try to circumvent regulations as much as they can. A notorious example is that of Volkswagen rigging their cars to activate emissions control only during testing in the laboratory setting. This was done so the cars would comply with the Clean Air Act, a regulation implemented in the USA to reduce air pollution.* The way the system is set up encourages companies to try and circumvent regulations to gain

---

* Here, "pollution" is used loosely, as it relates to the pollution threats we discussed earlier but also to ozone depletion, acidification, etc.

competitive advantage. Nevertheless, and in spite of the many flaws inherent in trying to implement a regulatory framework, regulations are crucial in the battle to make environmental sustainability a priority within our economic systems.

Ironically, the same competitive capitalist system that hinders the effect of regulation also benefits from it in some scenarios. A good example is the RoHS directive and how countries that wanted to do business with the EU had to adopt the restrictive practices imposed by the European economic bloc. Another example can be found in Brazil and many other countries, where corporations intending to develop a new site such as a factory or manufacturing plant need to submit their project to their country's environmental agency in order to receive the appropriate licenses and compliance papers before they can begin the development. Not having such authorization is very risky for big corporations, as they can end up paying big financial penalties due to noncompliance, and they are under far greater scrutiny than smaller players. For smaller companies, however, the risks are not usually great enough to justify the bureaucracy involved and possible delay. This is especially true in countries where the monitoring is weak. Left as is, only the larger companies would abide by the regulations imposed. However, it turns out that large corporations are big buyers of goods supplied by the smaller ones, which in turn are buyers of goods supplied by even smaller companies, and so on and so forth. It also turns out that the big ones are required to do business with companies that are compliant. So, to be able to compete for a big customer, small companies end up seeking appropriate permitting, which requires them to comply with regulations, and that trickles down to even smaller organizations in the supply chain. It is not a perfect system, but it assists with the crucial implementation of regulations and operates within the current economic setup we have globally. The bottom line here is that it is worthwhile to demand that the government impose regulations that push for sustainable products, supply chains, and services—and to elect leaders who will stand by their promises to do this.

Litigation

And when all else fails, we can always try to sue the government. No, seriously. There is even a term for it: environmental litigation. By many accounts, governments are failing to protect their citizens when they invest in or subsidize fossil fuel operations or when they do not properly regulate polluting activities. It is a violation of human rights, and it is especially detrimental to the younger citizens who are inheriting this planet. So far, nothing new, right? The novelty here is that this very argument has been holding up in court more and more around the world. There are examples of citizens from the Netherlands, Germany, and Pakistan suing the government and winning. Climate litigation is more present in the Global North, with the Global South only now starting to combat environmental degradation through legal action linking it to climate change.[208] The justice system has been increasingly ruling in favor of the people and demanding that the governments do more.[209] And it does not stop there—corporations have been losing cases, too, both around the human rights argument and the argument that corporations may be violating their duty of care. So, pressuring the government may mean going beyond asking, voting, and protesting; it may mean lawyering up and actually fighting in court.

This is a perfect example of collective little steps leading to big steps—people who share a common grievance banding together and using the court system as a way to effect systemic change. It's not individuals who are suing corporations and countries and winning, it's people joining together, often in partnership with nonprofit organizations. This is also where donating to certain charities can make a big difference—lawsuits are expensive in some countries, and the organizations supporting these litigation efforts are often largely dependent on donations.

Push for more efficient buildings

You may recall that in chapter 4 we talked about how about 7% of total GHG emissions arise from keeping places warm or cool. The good news is that we already know a lot about how to deal with this. The bad news is that people are not really aware of these solutions or how big the impact can be. And that is why we need amazing people like you to push for more efficient buildings.

Thermal comfort in the built environment is by no means new. The search for building techniques to keep indoor temperatures more consistent dates to thousands of years BC. The global oil crisis in the 1970s made energy efficiency a central point of concern, and soon building standards started taking this into account. In a broad sense, a building can be classified as standard energy, low energy, ultralow energy, or passive; the latter is when the energy requirements for heating or cooling are less than 15 kWh per meters squared of an area per year. For a sense of comparison, this is a tenfold reduction in respect to common old buildings, and a two to fourfold reduction in respect to current standards (both in terms of heating/cooling energy requirements and carbon dioxide emissions).[210] The challenge today is to not only optimize for thermal comfort, but to also do so in a way that minimizes material use and environmental impact.

The strategies for maximizing thermal comfort efficiency generally revolve around two basic building principles: passive design and envelope optimization. The former includes long-known approaches such as sun orientation (you know, facing north if you are in the Global South, or south if you are in the Global North), natural ventilation, and shading structures. The latter refers to the exterior of the building, i.e., the areas directly in contact with the outdoors. Here, a series of clever material selections can be a game changer in the temperature regulation of a building. This can be as simple as the use of insulation in the walls (keep the heat from entering in summer and escaping in winter), to more sophisticated ideas such as air tightness control (a great deal of energy is lost through air gaps in door/window frames), or the use of low-emissivity glass for windows. The latter is when the glass is coated with a substance that stops longwave radiation (which you might recall from previous chapters as heat energy) through the glass, while allowing shorter waves to pass (including visible light). The net effect is similar to that of the insulated wall (stop heat from entering in the summer and escaping in the winter), but with the added benefit that the shorter waves that do get in are then trapped inside as longwave radiation, warming the building from the inside.* So this kind of material application makes use of the greenhouse effect to the benefit of the building! These increased efficiencies mean that we don't need to rely as much on HVAC (heating, ventilation, and air conditioning) systems to cool down or warm up the built environment, meaning less energy is consumed and, therefore, less carbon is emitted.

The examples given thus far are by no means comprehensive, and the purpose of this chapter is not to explain them in depth, but rather to quickly overview some of the ways we can maximize thermal comfort and therefore minimize the energy requirements of the built environment. These technologies exist, and they are incredibly effective. So we

---

* You may be thinking, *Wait, won't that make the house too hot in the summer?* It is not an issue if the build is properly designed. The sun orientation changes during the summer and therefore wouldn't be hitting those windows. Moreover, there are simple mitigations like closing blinds to reduce energy coming in, or opening windows for ventilation. And there are also advanced mitigations like special coatings designed for different climates and smart windows that adjust the amount of energy that can come in based on the temperature indoors.

should be implementing these techniques everywhere, right? But are we? Unfortunately, no. Why?

The first reason is quite obvious: It costs more up front to apply these strategies. Even simple passive design like applying shading structures to block some heat requires having professionals capable of determining where the installations should be made, how many are required, what materials to build them with, etc. So, if you are an individual strapped for cash (most of us mortals) or a building company trying to make a profit (again, the vast majority of companies), it makes little sense to invest the extra dollars in something that, while important, a lot of people don't think about or appreciate. Of course, the very basics are generally applied to make a building "livable," but when it comes to those extra steps, it is a different story. This leads us straight into one of our core "what to do" items: awareness. We must understand that the built environment has a significant impact and that properly designing it to minimize energy requirements is way more effective than several of the little steps discussed earlier. Moreover, pushing for regulatory standards wherever you live can go a long way.

The truth is, these techniques cost more, and they require a delayed gratification mentality ("pay more now to save later") that humans are notoriously bad at. So one way to have energy efficiency implemented more broadly is through legislating the minimum standards that new buildings (and even renovations) must adhere to. Countries like Canada, Australia, the USA, Denmark, and Germany, to cite a few, all have their own set of regulations. Australia, which I am most familiar with, has a national energy rating system called (NatHERS), while each of the individual states and territories can have stricter minimum requirements that overrule the national one. Within a state, the individual local government areas (which you can think of as suburbs or neighborhoods) can set even stricter requirements. What this means is that your voice in your neighborhood is quite powerful, and your participation can go a long way! In practice, these standards mean that when you submit your project (whether it is for a new dwelling or a renovation) you will need to compute the expected thermal performance of the building, and it must pass the minimum standard before it is approved to commence. This standard is not exclusive to thermal comfort; it also includes other criteria such as water use and energy demand.

Here is the catch, though: Whenever we try to implement standards for things that will end up costing more due to regulations, people will try to circumvent those restrictions as much as they can, as I mentioned in a previous section about regulations. This is especially true if there is a lack of oversight, as people will be incentivized to do the bare minimum to avoid getting caught so they can maximize their profits (or minimize their losses). In theory, regulations work amazingly well, but in practice we see people make all kinds of maneuvers to ensure a project passes the minimum requirements, with those requirements often being ignored when the construction is actually being done because no one's following up to ensure they're being met. Too often we see the aesthetics of a building trumping the energy efficiency measures. The conversations follow the lines of:

"I want this façade to be covered in stunning big windows."

"Okay, that's going to require triple-glazed low-e glass."

"That is too expensive!"

"So why don't we decrease the number of windows?"

"You're killing my beautiful project!"

People just gotta have those big windows.

This goes back to the awareness thing. I believe that if property owners had a better understanding of the impact these strategies have in fighting climate change, many of these conversations would be different. Likewise, as companies raise the bar for their competitors in relation to how they design the buildings, others tend to follow suit. Maybe this is wishful thinking on my part, but there are some good examples that keep me hopeful (see some here[211]). Moreover, thermal comfort is not just about energy savings, it is also about, well, comfort.

Despite some of the problems around implementing these efficiency requirements, there are many success stories. This is as true for Australia[212] as it is for other countries. A case study in Cyprus identified that, after implementation of standards, buildings now consume up to 60% less energy compared to before the regulations were implemented.[213] Another study in Spain also demonstrated that advancements in the sector such as additional conditions and requirements were introduced.[214]

We have covered some of the strategies that can make buildings more efficient in respect to heating and cooling, but a lot of these approaches go way beyond heating and cooling. They also include strategies to save water, minimize lighting use, decrease the indirect impacts arising from the materials used in the construction of the buildings, and so on. Likewise, the carbon emission impact of buildings goes way beyond heating and cooling. If you classify the building sector as a whole, it is responsible for approximately 40% of the EU's energy consumption and 36% of its greenhouse gas emissions![215] Therefore, these initiatives can have an impact greater than the 7% from heating and cooling—it can tap into the 31% of making things, the 27% of electricity, and even the 19% of producing food.*[216,217]

In this context, it is also important to point out that urbanization and the increase in building stock are expected to rise in low-income countries like it has in the (already vastly urbanized) high-income countries, as there seems to be a strong direct correlation between urbanization and the income of a country.[218,219] And with that, we come back full circle to the idea of social progress as outlined in *Our Common Future* and that was discussed in the previous

---

* Check out an example of the latter in this video about an urban farm growing in (and on) an office building in Tokyo: *https://youtu.be/qJMZRIRkZWs*.

chapter—developing countries need to be given some leniency as they try to catch up with the living standards of more developed nations, and the countries that are already consuming and emitting more than their fair share need to do their part to become more energy efficient.

This chapter provides an overview of some of the big steps required to make the systemic changes needed to save our planet. While none of us as individuals can likely make these changes happen on our own, there is incredible power in organizing as groups to make our voices heard by the governments and corporations who are both doing the most harm and have the most power to stop it. I encourage you to do some research on how you can get involved in these environmental movements locally, nationally, and globally.

I'll be the first to admit this is hard, grueling work. Cultural shifts happen slowly, and government policy often takes even longer to catch up. Some of you may be thinking this all sounds way too complicated. Can't we just invent some new technology to fix all these problems? No, sorry. But allow me to explain the reasons for that in the next chapter.

CHAPTER 9

TECHNOLOGY TO SAVE US ALL?

F OR QUITE SOME time, the idea that the development of new technology can free us from the risk of an environmental disaster has been lingering in the minds of many. This idea is especially comforting and appealing since it theoretically allows us to keep our business-as-usual lifestyle without needing to cut back on any "luxuries." But how real is this silver bullet solution? This chapter highlights some of the technologies people are pinning their hopes on as potential solutions to our environmental problems. Many of them are still very theoretical at this point, and we'll explore how viable they may actually be. But first, let's investigate where we stand with Earth's resources and what it means—and doesn't mean—when we say they're being depleted.

NOT SO FAST...

Thomas Sowell, in his book *Basic Economics: A Common Sense Guide to the Economy*, explains that the limitation of oil (i.e., petroleum) we hear about is more of an economic limitation than a physical one. He argues that the oil resources are virtually inexhaustible because humans have consistently been able to come up with new ways to extract this resource from places that were once considered impossible to reach. He cites several reports that predicted the end of oil at various periods in time, all of which ended up being incorrect. In his words: "Many false predictions over the past century or more that we were 'running out' of various natural resources in a few years were based on confusing the economically available current supply at current prices with the ultimate physical supply in the earth, which is often vastly greater."[220] Specifically, the idea he tries to convey is that when a resource becomes scarce, its price goes up—which, in turn, allows processes that were once too expensive to be employed. This is a basic economic idea that is based on the interaction between supply and demand.

Similar arguments can be made for resources other than oil. For instance, once the price of water reaches a certain threshold, corporations will be incentivized to extract water from places that are currently difficult to reach (i.e., too expensive). Ethical considerations aside, I don't see any flaws in the reasoning around the economic model proposed here and don't have the knowledge to argue whether or not this will continue in the long run. However, the same argument Sowell uses in favor of the free market ensuring we won't run out of petroleum can be used against it. You see, if left only to the market to determine whether we should or should not use fossil fuels, the answer will always be that we should. As the bedrock of our energy matrix, fossil fuel extraction will remain profitable. The majority of our current infrastructure runs on fossil fuels, so it is currently cheaper and more convenient to use them. Other, cleaner sources of energy are precisely the opposite—more costly and inconvenient to handle. Solar is a good example of a technology that has been known for a long time, and it took the persistence of several pioneers who kept pushing the limit on efficiency and structure to allow society to have the current know-how and affordable prices. But despite this effort, the ability to completely switch the energy mix from fossil fuel power plants to solar still has several barriers and, most importantly, the required infrastructure is lacking.* At the same time, the big oil corporations have guaranteed oil reserves for the year of operation and for many years to come (what the industry calls "reserve-replacement ratio").[221,222] It takes a lot of political will and momentum to radically move away from fossil fuels. And you can be sure the oil behemoths will lobby and fight against this change to protect their enterprises (even if they publicly claim otherwise!).

In addition to the market pressure to extract more fossil fuels, there is another economic reality that poses a barrier to the "technology to save us all" idea. As technology progresses and new developments are achieved, the efficiency of certain processes tends to increase. This means that things cost less to run now than they did before (for example, less fuel per kilometer driven, less land per crop harvested, less energy consumed per processing of data). Theoretically, this should incur savings, which would incur a smaller environmental impact. However, human beings are funny animals. The moment something is cheaper (here, cheaper is meant in any terms you wish—costs less financially, takes less time, consumes less energy, etc.), we tend to consume more of it. So, if your car is more efficient, you end up driving more. If your farm is giving higher yields, you increase the size of your farm. If your computer is capable of undertaking more than one task at once, you multitask and open as many tabs as you can in your browser (yes, I see you). If electricity becomes cheaper, you use more of it and

---

* In the process of writing this book, solar became the cheapest form of energy, which is amazing! But the argument stands, as the infrastructure in place still favors the use of fossil fuels, and change is costly. Thus, the electrification movement we see (e.g., changing combustion engines into electric engines for cars) is a very expensive exercise. And it requires a big push from underdog players to make the shift happen. Moreover, solar is an intermittent source of energy and requires a way to store energy for when the sun is not shining. This takes the cost of solar energy up and often makes fossil fuels the default economic decision.

do not worry as much about wasting it as you did when it was an economic liability. So, after accounting for this extra consumption, are we left with an environmental impact that is smaller or greater than before? Short answer: As usual, it depends. A lot of times, it actually worsens the problem. This is called the Jevons paradox, after the English economist who first noticed the phenomenon in the nineteenth century. His work revolved around the use of coal to feed the steam engines that were generating an unprecedented energy output in England at the time. He talked about how much more efficient the engines were becoming and how that led to a greater consumption of coal instead of the opposite, which one might expect. In his words: "But such an improvement of the engine, when effected, will only accelerate anew the consumption of coal."[223]

So this is the paradox: A technological improvement that would be expected to result in a decrease in resource consumption ends up leading to an increase in the consumption of that very resource. This is not a rule, though. Sometimes the benefit of a given technological improvement outweighs the increase in consumption. In this case, the environmental impact is smaller than it would be had there not been any improvement, but nevertheless greater than it would be if the consumption hadn't increased due to that very improvement. Confusing? Let's throw a couple of numbers out there to make it simple and more tangible. Let's say you consume 40 liters of gasoline each week commuting with your vehicle (work, groceries, leisure, etc.). A good deal on a car comes along, so you exchange your vehicle for a more efficient one. Now your same routine consumes only 35 liters: less fuel consumption, less fuel production, less impact (disregard the impact of the vehicle exchange for now). But that extra amount of fuel left in your tank by the end of the week encourages you to take short trips to the beach every month or so. If these trips consume 9 liters per week, then the total difference (before and after the vehicle exchange) is 4 liters extra that you are consuming (more impact).

The Jevons paradox is good to keep in mind because it dissuades us of the idea that technology will be the silver bullet that can solve it all. Sure, I am a big fan of what new technologies can bring and what the development of science can do for us. Surely if we are to find a way out of the current environmental challenges, these developments will be a great part of it, but I believe they'll have to go hand in hand with drastic economic and policy changes (discussed in chapter 8). It is too easy to dismiss the problem because we believe someone else will come up with a magical solution that will make it all go away. It is easy because that means we don't have to do anything, just carry on with our lives the way we already do. You can look back 30 years and see for yourself the astonishing improvements we had in efficiency all across the board—in transportation, resource extraction, data processing, recycling, and more. Yet we keep increasing the distance traveled, the amount of resources consumed, the quantity of data, the sum of waste produced. Waiting for the problem to solve itself is kind of like walking toward the horizon, hoping to reach it eventually...which is perhaps even worse in some instances, given that with each step we take, the metaphorical horizon takes two or three steps backward.

Let us consider a concrete example that is closer to my field. I work with recycling, and in my early days viewing the world as a simple-to-solve place, I thought that if I could come up with better recycling processes, I could decrease the amount of mineral extraction (abiotic depletion through mining). It makes sense, right? If today we consume 10 units of iron worldwide per year and I am able to come up with a way to recycle 4 units per year, then next year we'd only need an additional 6 units. However, next year, if instead of consuming 10 units, we now consume 15 units, this means that the amount of iron mining has risen, albeit less than it would have had I not implemented my process. This situation gets tangled up with the Jevons paradox because, when recycling is taking place, mining all of a sudden seems more noble. So, unless we can reach an equilibrium situation in which all the materials required come from secondary sources (that is, recycled scrap), then the boost in recycling may not alleviate the resource depletion burden at all. Worse yet, perhaps it will push consumption forward because it brings the "it will be alright" fallacy with it.

I know many of you reading this book will still be betting a lot of your tickets on technology. There are indeed silver bullet promises such as geoengineering that we shall talk more about. I do not for a moment dismiss the importance of technology, nor do I think that new research and development does not have a big role to play in addressing the environmental challenges laid out in this book. In fact, I believe in these potentials so much that the remainder of this book is dedicated to looking into current technologies that can assist in tackling the issue. However, before we do start talking about the options, I would like to stress once again that technology is a tool we should use, but we can only truly solve the problems we face if such tools are leveraged with the other systemic changes required.

GEOENGINEERING

Okay, but what if we could somehow rectify the damage caused so far? You know, reverse the whole thing. A sort of get-out-of-jail-free card we could use if things start to spiral down fast (actually, faster). Is that even possible? When something sounds too good to be true, it probably is. But have you ever heard about geoengineering? ("Geo" as in "the Earth.") Simply put, it's about deliberate large-scale intervention to transform the planet's climate. There are multiple ideas about how exactly to do this. In general terms, they involve either sucking carbon dioxide out of the atmosphere and storing it elsewhere (like we'd suck an oil spill, for instance) or partially stopping the energy coming from the sun by reflecting more of it, like high-albedo surfaces do. We'll go over these ideas and how they would work (or not work) and then discuss how realistic it is to implement them.

Among the alternatives available, some are more mature technologically, some are more economically feasible, and some are more ambitious and dangerous. They all have pros and cons and will most likely work in conjunction with one another, instead of against each other; i.e., pitching them against each other is probably not the way to go. However, in a world with limited resources, the effort required to develop these different alternatives varies. And, obviously, a resource that goes into developing one technology is not going into developing an alternative technology. Thus, understanding both the advantages and shortfalls of each technology is important, as is pondering their known and unknown consequences.

Solar radiation management (aka, blocking the amount of incoming sunlight) gets a lot of press these days, so we'll dedicate a good portion of the geoengineering discussion to that. The idea of simply cooling down the atmosphere that we've been warming up seems like a simple enough solution. Yet other less flashy solutions can also be effective (and less dangerous)—from removing carbon from the atmosphere by using manmade equipment to using trees and allowing Mother Nature to do its thing.

With the goal of simplifying this part of the book in mind, let's split the technologies into two broad categories: negative emissions technologies (NETs) and solar radiation management (SRM). The former, as the name suggests, encompasses all the technologies that seek to remove carbon from the atmosphere, in an attempt to reverse the trajectory we've been seeing in the past century. Often the name carbon dioxide removal (CDR) is used instead of NET, but the terms mean the same thing. The latter, SRM, attempts to partially block the incoming radiation so the warming of the planet stops, or even drops. The idea is to regulate the temperature by "adjusting the tap of the hose" that sends us energy.

Negative emission technologies

There are several ways to remove carbon from the atmosphere. Some are as simple as planting more trees (but at a global scale that would actually influence Earth's climate and could then be considered geoengineering). Some are as complex as manipulating the chemical balance of the ocean to increase its carbon uptake. The NETs we'll cover in this chapter are carbon capture and sequestration (CCS), direct air capture (DAC), afforestation and reforestation (AR), and enhanced weathering (EW). However, as previously mentioned, there are plenty of others such as manipulating the carbon uptake of the ocean biologically using fertilizers, or chemically by changing the alkalinity; enhancing agriculture to increase the amount of carbon stored in soils; and converting biomass to biochar* so less carbon is released through bioenergy than is captured via photosynthesis.[224,225]

Carbon capture and sequestration

The first thing to acknowledge is that there are several names and variations within the carbon capture and sequestration (CCS) umbrella. Examples include carbon capture and storage (also CCS); carbon capture and utilization (CCU); a mix of both carbon capture and carbon utilization and storage (CCUS); and bioenergy with carbon capture and storage (BECCS), which I will cover briefly in this section, but there are also others. The main objective is always the same: Capture carbon emissions at the source (fuel combustion or industry), then transport it via ship or gas pipeline so it can either be used as a product or a service, or so it can be stored away from the atmosphere.

---

* Biochar is a type of charcoal that is produced by submitting organic materials, such as wood or plant waste, to a low-oxygen thermal process. You can think of biochar as a special kind of charcoal that's used in gardening and farming to improve soil health.

The number of CCUS facilities has been growing steadily; there are currently about 20 facilities operating worldwide, capturing about 130 million tons of carbon per year. This is only 0.25% of the 51 billion tons we currently emit globally, a very small contribution for now, but hopefully one that can have a greater impact as two things happen simultaneously: the operations increase (both because of higher efficiencies and a greater number of facilities) and the emissions decrease (taking the denominator down and the percentage figure up).[226]

Natural gas processing is the main source of carbon that CCS has been working with so far. Natural gas deposits generally contain huge amounts of CO_2, so its extraction is associated with significant releases of carbon. CCS can stop this carbon from going into the atmosphere and keep it stored instead. From a macro perspective, this approach makes no sense at all. We want to move away from fossil fuels, not extract more natural gas. As we extract more gas, we release more carbon (mind you, this is not even counting the burning of this gas later on!), and although CCS assists with making the situation better, ideally, CCS would not have to be used at all. Not because we should develop a way to extract natural gas without any release of CO_2, but because we should simply stop extracting gas, period.

Nonetheless, in addition to natural gas, the production of hydrogen, syngas,* fertilizer, and power generation are other examples of sources that can be (and are already partially) mitigated by CCS. And to a lesser degree, so too is the production of steel and bioethanol. In the transition to net zero emissions, having clean electricity and electrifying as much as possible is paramount. And, as

---

* Syngas, short for synthesis gas, is a mixture of gases—mainly hydrogen and carbon monoxide. You can think of it as a type of gas that can be used as a building block to create different types of fuel or chemicals, similar to how raw ingredients are used to make various food dishes. It's often used in industrial processes to produce things like synthetic fuels, fertilizers, and other chemicals.

we saw, making electricity is the second biggest source of carbon emissions globally. But we can't electrify every single thing. Some processes will still require fossil fuel (at least in the next few decades), and some processes emit carbon as a by-product of the chemical reactions required to obtain the product (producing cement is the classic example). And that is where end-of-pipe solutions (literally) such as CCS make the most sense. In other words, CCS can assist with the emissions that are unavoidable or technically difficult to abate.[227] Another important characteristic of CCS is that existing plants (power generating, industrial, etc.) can be retrofitted to utilize CCS so the infrastructure that is already in place can operate with a reduction in emissions.

As mentioned previously, once the carbon is captured there are different options as to what to do with it. The "storage" part in the CCS acronym implies storing it in deep geological formations, such as depleted oil and gas reservoirs and saline formations. If we go into its possible "uses" from the CCU or CCUS acronyms, then there is a big range of possibilities, some more realistic than others. The current main use, however, is pretty disheartening. The vast majority of captured CO_2 is sold to companies that use it to pump fossil fuels out of the ground. Yes, you heard me right. The main market for the captured carbon is for using it as a way to remove more carbon out of the ground. In very simple terms, the idea is to inject the gas through one hole, so that the oil is pushed up through another.

At least in theory, the idea is that there are other possible uses for the captured carbon, such as making fuels and chemicals (examples include ammonia and methanol), but in addition to these, all plastics are also made out of carbon. So making plastics and other products from captured carbon would allow for a somewhat circular economy: Carbon is captured and turned into fuels and products that release CO_2 when consumed, which is then captured and turned into these products again. This would be especially important for transportation methods that would be harder to electrify; planes being the best and most used example.* However, coming down from the cloud of idealism, what we find is that the fossil fuel extraction industry is the main financier of the CCUS initiatives so far, for better or worse.

The acronym we haven't covered yet is BECCS, which refers to bioenergy coupled with carbon storage. This is almost too good to be true: Grow crops (biomass) that will extract carbon from the atmosphere (through photosynthesis), convert the biomass into energy (combustion readily comes to mind, but it can be via fermentation, pyrolysis, etc.), then capture and store the carbon that would be released during this conversion process. Thus, instead of relying on the extraction of more fossil fuels to deal with the activities that will require, well, fuels, we could use this biomass that would also remove carbon from the atmosphere during its growth. The idea is indeed great, and many studies point to BECCS as paramount in achieving net zero and keeping warming below catastrophic levels. The downsides are mainly related to the land and water requirements for such endeavors, the latter being significantly greater than other competing technologies.[228] Note, also, that if we are capturing carbon but then releasing it again through activities such as combustion, this is not a carbon negative activity, but a carbon neutral one.

What we've learned so far is that the CCS technologies in general are great and can be a powerful tool in combating climate change. All good news, right? Well, that is what I thought until I started looking into it with more attention. You may recall that in chapter 6 we talked about how climate change is an important environmental threat, but not the only one. Then we introduced the LCA tool to assist with

---

* Batteries fall short on delivering energy with minimum weight like fossil fuels do. This can change with technological development, but as it stands, electrifying aviation seems to be far out into the future, at least for big commercial planes.

measuring other environmental impacts. Well, it turns out that while CCS is indeed great for capturing carbon, it has significant shortcomings. There are slight variations in the way carbon capture actually works, but the most common way is by grabbing the flue gas after combustion and reacting it with solvents/sorbents that selectively extract the CO_2 from the mixture, which is later separated from the solvent/sorbent. The now high-purity CO_2 can be transported to its final destination (be it for storage, usage, etc.). Among the sorbents used, MEA (monoethanolamine) has been traditionally employed. Don't linger on the fancy name, just note that it is an amine (nitrogen organic functional group). Here's the catch: LCA studies have shown that CCS using MEA scrubbing increases other environmental impacts, such as eutrophication and human toxicity, mainly due to the increase in NO_X emission.[229] This might be something we are willing to overlook while CCS operations are small, but if it were to increase (and if we are aiming to keep global warming under control, it should), then these other environmental impacts start to gain more relevance, and we'd be creating a big problem while trying to solve another big problem.

Direct air capture

As the name suggests, direct air capture (DAC) takes air from the atmosphere with the intention of separating carbon dioxide from the mix. It can be used to manage (or somewhat reverse) the emissions of carbon from scattered sources such as car exhaust or the organic decomposition of wastewater components. Individually, these are small sources of carbon emissions, but collectively they are significant, and it is impractical (and, currently, in some cases, impossible) to fit multiple carbon capture devices in each of these sources. Thus, DAC could come to the rescue. Moreover, it also has another important advantage over carbon capture: It is able to remove carbon that has already been emitted and is scattered throughout the atmosphere,* as opposed to concentrated in the outlet of a power plant or cement-making industry.

DAC works in a similar fashion to the mechanism of CCS that we just discussed, but with much smaller concentrations of carbon (CCS deals with concentrations from direct emissions in the order of 3%–15%, while DAC deals with the CO_2 concentration in the atmosphere, which is the 0.04% that we saw earlier in chapter 4). DAC works by reacting the captured air with substances that grab hold of the carbon dioxide, namely aqueous solutions of strong bases (e.g., sodium, potassium, or calcium hydroxide bases), amine adsorbents, or inorganic solid sorbents.[230] In addition to the reactants, DAC also requires heat to regenerate the medium; the most developed systems require temperatures above 800°C.[231] No wonder, then, that DAC costs more than CCS.

DAC costs between $100–$1000 per ton of carbon equivalent.[232] Back-of-the-napkin calculations would reveal that we "simply" need to invest anywhere from $5–$50 trillion to remove all the carbon in the atmosphere...per year.† However, the emissions associated with the whole DAC process would already revert some of the benefit. Examples of such emissions include the making of the required reagents (e.g., potassium hydroxide) plus the heat energy required (e.g., 800°C to regenerate the medium).[233] I am unsure how that equation plays out, but it would increase the total investment amount we just calculated. As if more complexity was needed, once a significant amount of carbon is removed from the atmosphere, Mother Nature will try to regain

---

* Because of this feature and because we have been failing over and over to reduce global emissions, many climate models show that keeping the average global temperature rise below 1.5°C is impossible without the use of DAC.

† This assumes 51 billion tons per year, which is the total amount of carbon equivalent being pumped yearly. This does not, however, take into account the emissions that are compensated by the reflection of solar radiation, as we discussed in chapter 4.

balance by releasing the excess carbon from the oceans to even things out (remember that the oceans act as a buffer capable of absorbing part of the excess carbon being emitted). And this release is not insignificant (it can be around 10%–20%[*][234])which brings up the cost of investment yet again. But guess what? Cost isn't the main hurdle to overcome. The main hurdle has to do with the rate of deployment and scaling up DAC enough to significantly address the issue. This requires not only the large capital investments already mentioned, but also technical improvements (making more of the reactants required, like hydroxides and amines), policy instruments, financial incentives, and a societal acceptance to give momentum to all of this.[235]

It is ironic that the cheapest way to operate a DAC plant is by powering it with natural gas (as opposed to renewables), which obviously has the side effect of pumping more carbon into the atmosphere (the rate at which DAC removes carbon is still greater than that of the natural gas side effect, if you are wondering). This is a good example of what we're going for when we say net zero. We don't expect to be able to reduce emissions to zero, i.e., not emit any greenhouse gas whatsoever, but we do want to emit so little that technologies such as CCS and DAC (and others we're about to see) can reduce the total ("net") emissions from reaching the atmosphere to zero.

The final caveat to all of this is that these projections assume emissions will stay as they are today. Best case scenario: They decrease over time and the total costs of DAC reduce with the decrease. Worst case scenario: We are so happy the problem is being solved that we turn up the engines (figuratively and literally) and increase the rate at which we pump carbon into the atmosphere. This is actually one of the main warnings concerning DAC. If deployed "correctly," it actually allows

---

* That is 10%–20% of carbon removed from the atmosphere. So for every 100 tons of CO_{2eq} we remove from the atmosphere, the oceans will release 10–20 tons back into the atmosphere.

emissions to increase a bit,[236] and this is arguably a problem because we will lose momentum to achieve the main objective: Decrease emissions. Exactly the same thing can be said of CCS and its variants.

Once the carbon has been removed from the atmosphere, the question becomes what to do with it. A handful of possibilities are currently on the table and more show up as people think of clever ways to utilize such carbon. Examples include using it to make clothes[237] and farming.[238] But the carbon can also be locked and stored away. This is direct air carbon capture and storage (DACCS), a variant of the mother technology. And if you've been paying attention, you'll note it is actually the same idea as CCS but with DAC added.

Afforestation and Reforestation

You may recall that in chapter 4 we talked about how trees are a great technology capable of absorbing carbon and releasing oxygen, with some additional perks on the side. Afforestation means growing forests where there were none, while reforestation is regrowing forests on land that used to have forests, but where they've been removed. These are usually considered in conjunction and abbreviated as AR (afforestation plus reforestation). Trees (and plants in general) are such well-designed beings that introducing them (or reintroducing them) is also considered NETs when done at scale (remember that these are *geo*engineering technologies). In fact, AR has the potential of being as effective as DAC or BECCS with the benefit of being economically cheaper (slightly cheaper than BECCS, and way cheaper than DAC). However, the Achilles' heel of AR is the amount of water and land it requires.[239]

In 2020, YouTube celebrities MrBeast and Mark Rober launched a campaign called #teamtrees to plant 20 million trees all over the world. At the time of writing, nine million trees had been planted, and funds were in place for 23 million trees. How impactful is such a feat? As is always the case, it is complicated. The actual analysis needs to take into account which trees were planted, where, how much land they are occupying, and a lot of other variables that researchers bundle together in complex models. For our purposes, we don't need to get that complicated. We can start by noting that a growing forest consumes about 0.3 tons of carbon per hectare per year.[240,241] If we then look at the spacing of plants directive from the UN Food and Agriculture Organization (FAO), we see that between 1,100 and 1,700 trees per hectare serves as a likely approximation.[*] So that means, given our rough numbers, that 23 million trees will remove about 5,000 tons of carbon per year, or about 0.00001% of the 51 billion tons we're dealing with. While 23 million trees is indeed a big number, the challenge we face is much bigger. And to be sure, planting lots of trees through collective action such as the one proposed by the YouTubers is great, but we need systemic changes to have the significant impact we require: lowering the emissions and implementing NETs systematically.

Obviously, the above numbers are a very rough estimate and a simplification of AR. If well-managed and deployed at scale by governments and agencies, AR with forest management can remove 3–18 billion tons of CO_2 per year, according to some very optimistic estimates.[242] The high end is about a third of the annual 51 billion tons of CO_2-equivalent emissions! More down-to-earth estimations of AR (pun intended) are on the low end, around three billion tons per year. But, of course, there are some important parameters to consider.

First, consider that the word "forest" is being used loosely here. Forests come in many shapes and forms and are ecosystems way

---

[*] Please note this is a gross approximation so we can relate the orders of magnitude; it is by no means a precise measure. The same applies to the number of trees per hectare that will pop up in the next few sentences.

greater than just their trees. A forest will only be a carbon sink if the absorption of carbon is greater than the release occurring at ground level, and this is not a given. The type of forest greatly changes the negative emission potential. Are we talking pine trees, eucalyptus, mangroves, or something else? It doesn't even have to be what you think of as a forest. Restoring peatland can have significant effects too. It can get even more complicated as trees vary widely not only in how fast they grow and how carbon dense they are, but also in a myriad of other things we generally don't think about. For instance, a darker tree will reflect less radiation than a lighter tree (or even no tree at all), and how much energy is absorbed is quite relevant if we intend on planting huge amounts of forest. Moreover, where will these forests be planted? In colder climates, where they will prevent the snow from settling on the ground? Remember that white surfaces like snow have a high albedo and can reflect a lot of incoming radiation. It actually turns out that AR in high latitudes may end up having a net warming effect![243] Conversely, in the tropics (low latitudes), trees grow faster, and this is often cited as the region in which AR would have the biggest impact.[244]

There is also the matter of space. How much land will be required so AR can effectively work as a NET? Estimates vary significantly and are dependent on another myriad of factors. One estimate calculates there is "room" for an additional 0.9 billion hectares of canopy cover (which could fit somewhere in the range of a trillion trees), an area a bit larger than the whole Brazilian territory; this would have the potential to store about a quarter of the total atmospheric carbon pool upon maturity.[245] In addition to the factors we already covered, there are other complex issues to consider, such as the cost of land itself and societal issues—who is willing to move or give up their land to make it available for AR (or, likely, who will be forced to move against their will)? Even when offered compensation or aid in moving, relocating communities is never a simple task.[*][246] Finally, there is the matter of time. AR takes a long time. It can be decades before a given region becomes carbon negative, and, as we've seen throughout this book, time is not something we can spare in dealing with the environmental crisis. In this sense, it is much more effective to keep the forests that already exist than to try to make up for deforestation with AR. This solution may seem obvious, but it is often the opposite of what has been happening. To make things simple, let us establish the hierarchy of action that should be taken: (1) preserve current forests while employing AR, (2) preserve current forests without AR, and (3) AR by itself.

One idea is to couple AR with BECCS, meaning the "forest" (again, loosely used) we plant would eventually be converted into energy, which would then be captured by CCS; then, the AR starts again. In addition to the problems of land use and water requirements mentioned for BECCS initiatives, it is important to note that the use of fertilizers can bring another environmental impact we've seen before: eutrophication. In a similar way that CCS assists with global warming yet increases human toxicity and eutrophication, the use of AR coupled with BECCS (or even just BECCS by itself) can assist with removing carbon from the atmosphere, but it is accompanied by a new environmental problem due to fertilizers running off and promoting eutrophication (see chapter 6 for a refresher).

---

* There is even an economic theory around what is called "stickiness," explained in the book *Good Economics for Hard Times* by MIT economists Abhijit V. Banerjee and Esther Duflo, which is worth reading if you are interested in this. In a nutshell, there is strong evidence that most human beings will not move even when there are better economic prospects elsewhere. People tend to "stick" to where they are and to where they know.

Enhanced weathering

I recently visited a place called Moon Valley (Vale da Lua, Chapada dos Veadeiros) in Brazil. It receives its name because the rock surface resembles that of the moon (it is beautiful—look it up). Interestingly, the process that makes the surface look like Swiss cheese also assists in removing carbon dioxide from the atmosphere.

Weathering, at its core, is the process of breaking rocks into smaller pieces. Not so much like a backhoe breaking big rocks in half, but more like a jeweler slowly chipping away tiny bits and corners. This is a natural phenomenon and is part of the Earth's rock cycle. What does this have to do with climate change and the other environmental impacts? I'm glad you asked. Rocks (minerals) are made up of compounds that are slowly dissolved by water. Calcite ($CaCO_3$) and silicate are classic examples, and they react with water and carbon dioxide to form bicarbonates (HCO_3^-) that are carried away into the oceans, effectively trapping the carbon (look at the bicarbonate formula again to see there is a CO_2 in there) in the ocean floor or dissolved in the ocean itself.

To be sure, this is a very slow process. It is, after all, a geological process. But it is happening all around the world at every instant. Natural weathering is capable of absorbing 0.3%–0.5% of the global fossil fuel emissions.[247,248] This is not much and won't make a dent in the problem at hand, but here comes the enhancement part in *enhanced* weathering: What if we accelerated this process so more carbon is absorbed? How much faster can we make this phenomenon happen, and how much carbon can it remove?

One way to enhance the weathering is to crush (or pulverize) rocks into fine particles and spread them over a large area. Farmlands are particularly good spots for this because not only is there plenty of area available, but also because the "fine rocks" also benefit the crops: This process releases nutrients (e.g., phosphorus), makes the soil less acidic, and stabilizes soil organic matter.[249] It makes so much sense that farmers have been doing it for centuries, long before the scientific principles at play were fully understood. So it is a win-win situation. Remember that a large area is beneficial because the more land, the more contact area between the rock and the air for the reaction to take place. The same idea applies to the size of the crushed rocks themselves. The smaller the rock, the greater the total surface area, and the faster the CO_2 removal occurs. Interestingly, though, we need to think about how we are going to crush the rocks in the first place. The energy required for this can negate 10%–30% of the benefit, depending on the source of energy.[250]

If you are thinking ahead, you probably noted that enhanced weathering can be paired with forestry (AR) to replace cropland, so a trifecta can be achieved: the removal of CO_2 through enhanced weathering, increased fertilization of the land, and growth of forests that remove (trap) even more CO_2. In fact, the optimum effect can be achieved by applying the pulverized rocks to tropical forests, which can achieve about 80%–89% of the effect while requiring only a third of the land and 70% of the rock mass.[251]

I've been using the term "rock" here loosely (so loosely as to make any geologist cringe). But enhanced weathering can also be achieved by weathering materials produced from the manufacturing of steel, aluminum, cement, lime, nickel and other important industrial processes. Things like steel slag, cement kiln dust, concrete in building products and demolition waste, and fuel ashes are often rich in silicates and hydroxide minerals that can be "enhanced weathered" and will capture carbon in the process. It is a way to make use of current "waste" produced by industrial activity while combating climate change. The magnitude of this is not that significant today, but if we start to use such materials now (and get over the learning curve of doing so), they may contribute significantly in a couple of decades.[252]

Now picture this: The "rock" has reacted with water and carbon dioxide and has created bicarbonate ions. These ions are dissolved in water and will go wherever the water is flowing, so they can end up in rivers, which can end up in the ocean. The result is that the sequestered carbon can effectively be stored (locked away from the atmosphere) for a long time in the ocean (>100,000 years). But there is an additional benefit: Bicarbonate will increase the alkalinity of the oceans. So, enhanced weathering can assist not only with the carbon accumulation in the atmosphere (and all its consequences), but it can also counter ocean acidification. Better yet, if deployed first on land (as in the case of farmland and forests), it can counter terrestrial acidification (as you may recall from chapter 6). Obviously this only works if the deployment is occurring on land that is more acidic in the first place.

On the other hand, whenever we hear that a process releases "nutrients," we should think of eutrophication. Indeed, our current understanding of what would happen to aquatic systems, and the effects on air, water, and soil pollution, is still rudimentary. Another source of concern is the fact that the application of fine dust particles can have negative health effects. Refer back to the air pollution section in chapter 6, and you'll recall that inhaling small particles has detrimental effects. So choosing where to deploy large quantities of fine "rock dust" is more critical than a first assessment may suggest.[253] Finally, as was the case with DAC and AR, nature will try to balance things out. So removing one ton of CO_2 from the atmosphere will incur a rebound effect of emissions from natural carbon sinks like the ocean, and less than one ton will effectively have been removed.[254] And just like the case with AR, depending on where you deploy these particles, there may be changes in the surface's albedo that counter the cooling effect we are pursuing with these tactics in the first place.[255]

In spite of these caveats, it still sounds like a great plan. But how much carbon can EW actually remove? A study calculates that the application of 275 Gt (gigatons)* of basalt dust can remove about 2.5 billion metric tons of carbon per year, so effectively about 5% of our current emission of 51 billion metric tons. Not a huge sequestration, but one that could assist nonetheless.[256]

Note, too, that so far, all the options discussed have potential, but they also come with many caveats and limitations. None is the ultimate solution to climate change, nor do any of them completely allow us to bypass the other big steps from chapter 8.

Let's now look at the craziest idea of all. Instead of trying to fix our mess by sucking the excess carbon we released out of the atmosphere, let's simply block the incoming radiation from the sun to try and balance things out. What could go wrong?

Solar radiation management

Solar radiation management (or solar geoengineering) is the name given to the idea of stopping the incoming radiation from the sun. That is, stop the radiation from getting stuck inside our atmosphere in the first place. Not all of it, of course, but part of it. That's where the "management" part of the name comes in. In theory, we could calculate the amount of radiation that would still allow nature to function normally but would prevent the planet from heating further. Tempting, isn't it? In chapter 4, we talked about natural counterforces to global warming and how substances like sulfates have this effect. Therefore, if we can release great amounts of sulfur into the atmosphere† in a controlled fashion to form stratospheric sulfate aerosols, we can adjust the amount of radiation being reflected to our liking

---

* A gigaton is one billion metric tons, or one trillion kg, or 2.2 trillion pounds.

† It can be in the troposphere or the stratosphere, the latter being preferred because the reflective particles last longer (longer residence time) than in the former.

(again, theoretically). And that is precisely the idea. The "how to do it" has different propositions, from big hoses placed in the sky to the use of aircrafts capable of releasing the necessary quantities of the substance required. Engineering how to do it is not a deterrent, as there are actually some tested concepts for how to deploy it. Nor is the price to get such an endeavor on its way a deterrent. In terms of financing, this may actually be one of the cheapest solutions to global warming problems.[257] At least at first. But let's look at the dark side of solar radiation management.

I've spoken about uncertainty time and time again in this book. Perhaps solar radiation management has the biggest uncertainty of all: What will be the full scale of consequences of implementing such a "solution"? You see, in science, after a hypothesis is proposed, it is tested through the use of experiments. Similarly, in engineering, we'll first create a proof-of-concept prototype, test it for safety and efficacy, and enhance it as necessary. Only after these tests, when the product is deemed acceptable, do we release it. Even when delivering the famous minimum viable product (MVP), it still needs to be…well, viable. The problem with solar radiation management is that there is no way for us to test it on a small scale, gather data, and then fix any problems before deploying it globally. Local tests will either not be big enough to achieve the desired result or won't be representative of all the interactions the full deployment would entail. So if we happen to go ahead with it, then everyone (literally) on Earth is going ahead with it.[258,259] Hopefully, you can see not only the safety challenge here, but also the ethical and political dilemmas (even if the majority is in favor, can it be imposed on the rest, even if it threatens their lives?).

At this point, you may be thinking, *Okay, we can't realistically test it before deploying it globally; I get that, but can't we model the consequences? After all, climate science also can't accurately predict everything, but we can still produce models to assist us in seeing what the big trends are, right?* Right! We can. And some projections show that solar radiation management would disrupt the summer monsoons, which in itself is bad, given that so many people rely on monsoon precipitation each year to cultivate food,[260] but it could also trickle down into even more damaging consequences. The Amazon forest could be severely affected by a significant reduction in precipitation. Since solar radiation management is basically pumping the same stuff into the atmosphere that gets emitted during volcanic eruptions, we can look to big eruptions of the past as models of potential negative consequences. There appears to be some correlation with severe drought, but causal links (as opposed to correlations) cannot be established.[261] And here the feedback loop strikes again, bringing a whole lot of uncertainty with it. Would a significant reduction in the Amazon forest precipitation start a chain effect in which the forest cannot sustain itself any longer, or will it simply be a one-time thing? Perhaps the drying up of the Amazon forest's floating rivers* would severely change the ecosystem in the whole macroregion, which would then mean that surrounding areas would be affected, and all of a sudden, we have a whole new world. What would it look like?

Okay, so uncertainty about the consequences is one of the main issues with the solar radiation management idea. But there is more. Again, looking at projection models, it seems that while average temperatures could indeed remain constant, that would be at the expense of some regions warming up faster while other regions cool significantly. Picture your kitchen on a very hot day. You are sweating heaps and feeling ill. But someone comes and says that the average temperature of the kitchen is not that bad if you combine

---

* Floating rivers is a term coined by Brazilian scientist Evaristo Eduardo de Miranda. The Amazon floating rivers refers to the vast volumes of water vapor that flow above the Amazon rainforest, transported by the region's dense vegetation through evapotranspiration. They play a crucial role in distributing moisture and rainfall across South America, significantly influencing the climate and ecosystems of the entire region.

the temperature of the freezer, the fridge, and the room...as silly as that might sound, it is kind of the same deal. Sure, there will be Goldilocks zones where the temperature is just right (perhaps if you open the fridge and stand next to it), but you wouldn't want to be away from the fridge or inside your freezer... And here another dilemma enters the scene: Which regions would warm up and which would cool down? If we could choose by country, how would this be chosen? Geoengineering interventions in the Northern Hemisphere are projected to cause droughts in the Sahelian region (the region in Africa that includes Niger, Mali, Burkina Faso, Senegal, Chad, and the Sudan). If deployed in the Southern Hemisphere, it may actually increase the region's vegetation, but upset the rainfall pattern in the northeast region of Brazil.[262] Which country will step up and say they are happy to suffer the modeled negative consequences so others can benefit? Do you see any government doing this any time in the future? Or ever?

Yet another problem of trying to partially block the sun's radiation is that we may be enslaved by this very act. If we decide to go ahead with this idea, we better stop emitting GHGs into the atmosphere. Why? SRM masks the warming effect of GHGs by reflecting some solar radiation away from Earth, but it doesn't reduce the GHGs themselves. If we rely on SRM without cutting GHG emissions, we become dependent on it to maintain stable temperatures. If SRM were suddenly stopped, the accumulated GHGs would cause a rapid and extreme increase in global temperatures, leading to severe climate impacts. Thus, the relationship between SRM and GHG emissions is critical: Without reducing emissions, SRM could trap us in a situation in which stopping it would trigger unprecedented warming. If we decide to stop and revert back to the natural amount of radiation, the rate of heating and droughts would be unprecedented.[263,264,265,266] So we'd pretty much be the hostages of our own solution. Once we start, we can't stop.

"Wait," you may say, "But that's assuming we don't stop emitting GHGs, right? Why wouldn't we stop?"

Well, have we stopped so far? Despite all the studies and scientific warnings about the subject, we aren't really keeping up with the required changes, are we? So, there is a strong argument from those who believe that deploying solar radiation management to "buy us time" would turn out to be an encouragement to maintain business as usual. There is no way of knowing for sure, but if there is one thing we know about human nature, it is that we love to postpone action. I personally don't think it would be much different in this case. Especially when action means challenging the status quo that benefits powerful corporations and enterprises.

Interestingly, reverting to solar radiation management would likely reduce the efficiency of solar panels. At the time of writing, solar still represents, globally, a very modest source of electricity generation, but it has been growing rapidly. As its uptake accelerates, so does the increase in other renewable energy sources like wind, replacing fossil fuel energy sources (e.g., coal power plants). Picture, however, the irony of having to produce more energy with coal power plants to make up for the energy loss from solar because we pumped sulfur into the atmosphere to reduce the effects of global warming.

Last but certainly not least on our "dark side" list is the fact that geoengineering is aimed at one problem and one problem only: global warming. Throughout this book, we looked at how the environmental challenges are more complex than one single issue and how there are several different areas that need remediation. If we briefly ignore the heavy clouds of uncertainty around the idea of adopting solar radiation management and assume it would work like a charm to lower the globe's average temperature, we would be left with several *other* problems. Some of which, mind you, could be made worse due to the very sulfate deployment meant to save the environment. The problems include extensive depletion of the Artic ozone layer due to the presence of both sulfate (with high surface-area density from aerosols) and cold conditions in the polar stratosphere.[267] There's also the threatening of biodiversity, since solar geoengineering changes the climate rapidly, while species tend to adapt slowly. We're already seeing a biodiversity threat with species going extinct due to climate change (as we covered in chapter 6), so rapidly terminating solar geoengineering at any point would significantly exacerbate this threat.[268] Another problem is the increase in ocean acidification because a cooler atmosphere would increase the ocean CO_2 uptake[269] (which, mind you, can be exacerbated even more if we indeed continue to emit carbon into the atmosphere while engaging with geoengineering).

Geoengineering is so tempting (again, a get-out-of-jail-free card!) that we really do not want to let go of it. We keep hoping it is a valid option, so we keep researching it and finding ways to convince ourselves it will work. But because it is so appealing, it is also dangerous: Our bias for wanting it to work can blind us. This circles back to Leonard Mlodinow's book *Subliminal*, mentioned in chapter 1: Our mind is very good at believing that the things we want are good for us. It is much like the popularity of research that says chocolate, coffee, or wine are good for our health. We all want them to be! But this desire for something to be true is often a sign that we should be extra skeptical about it. With that said, we have only scratched the surface of this topic. There are several books about various geoengineering ideas, detailed analyses, cost assessments, deployment ideas, and pros and cons.

While I have laid out several reasons why geoengineering may turn out to be very dangerous, I would not discard it from our array of options just yet. It may turn out to be the best shot we have after a succession of failed attempts at reducing our GHG emissions. But we definitely need to decrease the uncertainty around it through further research,[270] and possibly through the development of enhanced techniques. In the case of solar radiation management, the climate system seems to respond quickly to the reduction in solar radiation, which means that it is probably better to delay its deployment until the climate situation is unmanageable instead of deploying it early, so the potential negative effects will pale in comparison to what is already being felt.[271] As it stands, I am personally against this idea, but then again, as it stands, we are not going in the right direction as fast as we should.

While the ideas in this chapter have the potential to help mitigate the climate catastrophe, I hope I have convinced you that none of them are silver bullet solutions that will allow us to maintain business as usual in terms of fossil fuel consumption, carbon emissions, and other status quo economic activities for that matter. For any of these technologies to be truly effective, they must be done in tandem with coordinated global efforts to make meaningful systemic changes in how we consume energy and how we structure our institutions and our economy to function in equilibrium with nature.

FINAL THOUGHTS

W E'VE COME A long way in this book, starting with the origins of human-caused environmental impacts, then exploring the sources of climate change and other environmental threats and diving into the political and economic structures that exist and how they affect our ability to respond to the environmental crisis. We've talked about how climate change is one among several other environmental threats, and we've discussed these other threats in more detail. We've discussed tools that can be used to navigate the ocean of information (and misinformation) out there, what we can do as individuals to fight for a sustainable future, the big systemic changes that need to happen, and some of the novel technologies being developed to mitigate the effects of climate change.

This book was written to reach an audience beyond university walls, in a relatable way that is (hopefully) easy to understand, and to try to simplify a complex problem while remaining true to current scientific evidence and consensus. The objective is that, upon finishing this book, you'll have an understanding of the major environmental issues facing our planet, their importance, and what is needed to address the situation. Specifically, this book should have assisted you in developing your own environmental lens to evaluate different situations and determine what you can do on your own. After reading this book, I also hope you agree with me that while an individual's contribution is of utmost importance, the change we really need requires collective action and a drastic systemic shift. That is, it is not enough to address your own personal impact without addressing the global impact too.

This understanding is of the utmost importance, yet it is also too easy to misinterpret. In *The Madhouse Effect: How Climate Change Denial Is Threatening Our Planet, Destroying Our Politics, and Driving Us Crazy*, author Michael E. Mann and political cartoonist Tom Toles perfectly sum up the issue in the following sentence: "Ultimately, we are left with one real solution: reducing our collective carbon footprint."[272] However, note how easy it is to interpret "reducing our collective carbon footprint" as "we *each* reduce *our own* carbon footprint," effectively ignoring the "collective" part of the statement. Rather, we should read this as "we *all* work to reduce the *collective* carbon footprint." Slight change in interpretation, huge difference in the course of action to be taken!

Most of us go about life without really thinking about our obligations to the wider society. But as human beings who share a single planet and who do not live completely isolated, we have moral and practical duties to serve. The extent of these duties has been widely discussed by different schools of thought and philosophies (e.g., libertarians, Kantians, utilitarians, etc.), and it is not within the scope of this book to debate these. But as a scientist and engineer—and, most importantly, as a citizen of this planet—it seems quite obvious that we all have a practical obligation toward the environment for the simple reason that ignoring this obligation up till now has led us to where we are. If you live in a civilized society, you respect the laws (both legal and social) that were agreed upon and that theoretically allow you to live alongside others. If you do not agree with such laws, you can take action to change them. While the power structure is admittedly not fair to everyone, this is the agreement we're all subject to in modern democracies. So the deal you enter in order to have a paved road, take public transportation, receive potable water at home, have access to public goods and a sewage system, etc., is that you will pay your taxes, among other duties. By the same token, I would argue, if you want to live on planet Earth, you must live a sustainable life and have a sustainable environmental footprint.

In the 2016 movie *Captain Fantastic* (which is *not* a superhero movie, at least not in the way most people picture superheroes, anyway), a couple abandons urban life to live off-grid with their kids in the forest in a way that is almost completely isolated from society—they build their own shelter using material they gather on their own, grow and hunt their own food, etc. They've chosen a life in which they do not owe anything to society and society does not owe them anything in return. I cannot calculate their carbon footprint, their waste generation rate, or any other metric to ensure they were living a sustainable lifestyle as a family, but let's assume they were (I suspect their environmental impact, as portrayed in the movie, was negligible, but let's just leave it as sustainable) for the sake of the following question: Would that form of living solve our current environmental problems? The short answer is no. First, because not everyone can live like the family in the movie. We can't all go back to being hunter-gatherers. Not because most of us would be unfit and not up to the task, but because the world in the twenty-first century simply cannot bear the weight of or supply the resources for eight billion–plus homo sapiens living as hunter-gatherers. There is not enough land; there is not enough food.

But perhaps you are enticed by the idea of living isolated, with a negligible impact—if you decide to do it, would your individual action solve the problem? Again, no. It would assist, sure. You'd be leaving a smaller environmental footprint, which is great. But the root causes of the problem would still exist—business and industry as usual, agriculture as usual, government as usual, society as usual. The root causes of the problem would remain unchanged and unchallenged even if a large proportion of the global population were to live in their own eco-friendly retreat. Let us try to paint a better mental picture: Suppose half of the world's population magically started to live with a negligible environmental footprint. How would that affect the current situation?

Give it a thought. Would it be enough to allow business as usual for everyone else?

What if I told you that the half that magically became eco-friendly are among the poorest of the world? How does that affect your first conclusion? Note that, as an example, the OECD countries[*] represent only 18% of the global population (2019 data[273]) but emit about 35% of the global CO_2 (2016 data[274]). Another example is energy consumption worldwide. A decent living standard can require as little as 15.3 GJ (gigajoules) per person per year, but current energy consumption ranges from under 5 GJ for people in the poorest nations to over 200 GJ on the other end of the spectrum.[275] My point is that the magical scenario of half the world living in an eco-friendly manner is not far-fetched; it is actually close to the current global situation. The problem is the other half, those of us who are making the most environmental impacts (and yes, I do include you in this by virtue of you having the leisure and means to even read this book) are the ones whose responsibility it is to make the biggest changes. And as I've mentioned throughout this book, it's not enough to individually contribute by lowering your own impact; we should all strive to change the collective sphere as well, transforming the status quo and how the global economy currently runs without regard for its negative environmental effects.

To change the collective sphere and transform the status quo, we must look beyond our individual contributions. That means each of us using the resources and privileges available to us and putting them to good use for a good cause. I wrote this book because I wanted to

---

[*] OECD stands for the Organization for Economic Cooperation and Development, which is a group comprising what are typically thought of as high-income countries, or developed nations. Though the OECD itself stands for something else, taking a sample from OECD countries is a proxy for taking a sample of high-income countries. Australia, Belgium, Denmark, France, Netherlands, Sweden, UK, and US are examples of member countries.

share my experiences and knowledge, and I thought I could organize the scattered information out there about these subjects in a way that is easy to digest. I hope I have delivered on that goal.

This desire to use the tools available to me to make an impact is also what led me to try to change a local law in a city I lived in previously, and why I have engaged with local and federal environmental protection agencies to provide information on how to improve regulations. It is also what drives me to work on research, trying to separate the good actions (the ones that really work and can make a difference) from the rest. I was fortunate enough to receive an amazing education in a country that is very unequal in its distribution of wealth and other resources. I learned English at a very early age, have always had food on my plate, never had to sleep without shelter, have had a family that stimulated me intellectually, and have always had unfair advantages that pushed me forward.

So I ask you: What are your advantages and how can you use them toward creating a more sustainable life on this planet—for yourself, others, and the world as a whole? What are your resources? Money? Time? Your network? Whatever privileges you have, figure out what they are and put them to good use.

As a rule of thumb, create a practice of making decisions using an ethical and collective framework. When in doubt, ask yourself: What would happen if everyone did the same thing I am doing? Would that be sustainable? It is a good question to ask ourselves every so often, especially because the answer will vary over time, as our habits change, as we organize ourselves as groups, and as technology evolves. If you are not satisfied with your answer, think about what you can do to change. Revisit the chapter on little steps to look for ideas and inspiration, but always keep in mind the big picture, i.e., don't get caught up by only your individual impact.

You might wonder what exactly would be needed in terms of "economic change" for us all to live sustainably. Thinking about the off-grid living in *Captain Fantastic*, how far back would we need to go technologically to live a sustainable life? How much regress is required? How much of our current lifestyle needs to be dialed back? What would a society like that even look like? To partially answer these questions, I would like to share the bottom-up modeling of a recent study that tried to answer how much energy per capita would be required to live a decent life for the projected global population of 2050. The study finds that even with a ~30% increase in population, energy consumption could be 60% lower than today's consumption, and that for the current highest per-capita consumers, energy consumption could be cut by roughly 95% and still allow for decent living standards. When I hear this kind of information, I am always skeptical, thinking *Yeah, right. I can only imagine what* decent *standard means*. But when reading the model in detail, I found the decent standard proposed was well beyond my expectations and doesn't require "going back in time" at all. The model's concept of "decent" includes energy required for nutrition (i.e., producing, supplying, cooking, and storing food); shelter and living conditions (e.g., lighting, heating); hygiene (including water and waste services); clothing; healthcare; education; communications (phone, computer, data); and mobility (vehicle production, propulsion, and required infrastructure). It's worth noting that nutrition and mobility require the highest share of energy among the categories.[276]

As a response to those who claim that environmentalists are proposing we need to return to living in caves to have a sustainable lifestyle, the authors show that such "caves" would have

> highly-efficient facilities for cooking, storing food and washing clothes; low-energy lighting throughout; 50 L of clean water supplied per day per person, with 15 L heated to a comfortable bathing temperature; they maintain an air temperature of around 20°C throughout the year, irrespective of geography; have a computer with access to global ICT networks; are linked

to extensive transport networks providing ~5,000–15,000 km of mobility per person each year via various modes; and are also served by substantially larger caves where universal healthcare is available and others that provide education for everyone between 5 and 19 years old.[277]

So hopefully we've settled the matter of how much we'd need to go back in time to live more in harmony with the environment. Sure, there will be some cutting back, but it is not as bad as one might think. And we can get there through all of us taking the little steps to start.

I see the little steps as something to walk toward slowly. Changing human habits requires a great amount of effort and usually doesn't work when forced upon individuals. I truly believe we'll fail if we focus all our efforts on trying to change the habits of people who are currently living beyond their "individual environmental budget." That does not mean that no change is needed, or that we shouldn't push for educating people about their impacts and how to live differently. But I think this change needs to come gradually and smartly. We should strive to lower the impact of the most affluent while allowing the (somewhat controlled) increase in the impact of the others—one generation at a time. On the other hand, the big steps are needed *now* in the form of systemic hard turns that require immense collective action and power. The truth is, revolutions don't require 100% of people to become revolutionaries; they need coordinated action and momentum. While a daunting task, I do believe systemic change is easier to achieve than waiting for the whole world population to change its habits. Groups of engaged people leading the way can turn the tide to benefit everyone.

And just like the action of reducing an individual's share of emissions won't be enough to change course, neither will reducing the emissions of a single country. Even if the largest polluter today, China, was to reduce its emissions by a staggering 80%, we'd still be far from achieving the current target. It must be a globally coordinated action.

In these final thoughts, I'd like to leave you with some key ideas: We, as a global society, are not living sustainably. We are causing damage through the consumption of natural resources and the emissions caused by our activities (waste, gases, effluents, etc.) beyond the rate the Earth can replenish. A critical issue resulting from this uncontrolled release of emissions is climate change, a paramount issue that needs to be addressed now, but it is only one of many environmental threats. Anthropogenic climate change is real and there is a scientific consensus about it, yet there are also several degrees of uncertainty around the consequences and mechanisms at play—and there always will be, as long as we are using the scientific method. It is important to understand the big picture and to differentiate between what is based on a reliable body of evidence and what is speculation. Life cycle assessment is a tool that can assist in this global understanding, and in quantifying and evaluating environmental impacts in many forms. Systemic change is required, and it can be achieved through the big steps, which, sure, involve your role as a consumer, but also involve so much more than that. Your role in the big steps can include pressuring the government (local, state, federal) and corporations, becoming an activist, supporting action passively through donations, educating people, and many other approaches. The little steps are important, but be wise to not get stuck there. We are targeting climate change for this systemic change, but keep an eye out for the other environmental threats and apply the same lessons to those challenges.

One last thought I'd love for you to take from the book is the following: The Earth will survive. It will find a way to adapt. Nature will eventually find equilibrium no matter what humans throw at it, and it will find a way to thrive. The real uncertainty is whether humans will be part of this next equilibrium. In the event of catastrophic consequences, many species will die (many are already gone!), and only the fittest will survive. Will humans be among the fittest? And if so, which humans? Only the most affluent? And under which conditions? Perhaps our diets and housing will have to change drastically, and speaking even more long term, even biological changes like our body temperature and breathing mechanisms, as well. And if "we" change so much, would it still be correct to consider this species as "we"? The forms in which humankind organizes itself will surely have to change. The landscape will have changed. Political divisions will take new forms; many cities that exist today will be gone. What will society look like in the future—if there is a society at all?

It is incorrect to think that humans need to protect the environment for Earth's sake, for it will continue regardless of how much we destroy. We have to worry about protecting ourselves—we need to allow Earth to protect us. We need to allow nature to keep the conditions (indeed, very specific conditions) that permit human beings to survive.

The United Nations' 1987 Brundtland Report, or *Our Common Future*, defined sustainability as "development that meets the needs of the present without compromising the ability of future generations to meet their own needs." In 2015, the UN expanded on this by putting forward their sustainable development goals (SDGs) to reach by 2030, which aim to allow for a "better and more sustainable future for all."[278] These goals span across a broad range of topics, many of which have been covered in this book: sustainable consumption and production, climate action, oceans, biodiversity, forests, desertification, and to a somewhat lesser extent, energy, infrastructure, industrialization, cities, and health. But the SDGs go beyond the environment and include important goals like peace, justice, building strong institutions, ending poverty in all its forms, and gender equality and women's empowerment.

As you can see, the UN views sustainable development as encompassing more than just the environment. Environmental justice and social justice are intertwined and will require the same profound structural change to address the problems and challenges that go beyond climate change.

I want to end on a hopeful note, not just so you feel inspired instead of discouraged when you close this book, but also because I would not have written it if I did not believe we could overcome this challenge. While change is not happening nearly as fast as we need it to, there are many signs that we are moving in the right direction. In many places in the world, for the first time, renewable energy sources like wind and solar are generating more electricity than fossil fuels. Dozens of cities, states, and entire nations (such as South Australia, Costa Rica, and Norway) have managed to run on 100% renewable energy for varying periods of time. The renewable energy industry continues to see exponential growth year after year and is on track to surpass coal to become the largest source of electricity generation in the world. Younger generations have grown up with an understanding of climate change and how important it is to address it—and they are fed up with the previous generations for dragging their feet. Studies show that Gen Z and millennials are more concerned about climate change than previous generations, and they're also more engaged in action. Studies also show that people increasingly care about nature loss and biodiversity issues, especially in developing nations where these issues are felt more acutely; markets are changing as people demand more sustainable products, and global news coverage of environmental issues is increasing. Voices around the world are demanding that governments and corporations take the brave, bold, and big steps necessary to reverse course on environmental destruction. Now it's time for your voice to join them.

ACKNOWLEDGMENTS

First and foremost, I would like to thank my loving wife, Adriane, who has always been supportive in every facet of our lives. She is a fierce climate advocate with whom I have shared many discussions, and she has shaped this book in many ways.

I would like to thank my eternal professor, Dr. Andrea Moura Bernardes, who has been an inspiration to me since the first class I took with her and continues to be, even after all these years. Many of the ideas in this book are either lessons learned from her or inspired by her thinking and teaching.

I thank my family for providing me with everything I needed to pursue my studies and acquire the knowledge I share with you today, and for creating a fertile environment for my growth and learning. My supportive family is central to who I am and the accomplishments I've achieved.

I am grateful to Dr. Andrea ML Ribeiro, aka Mom, who has not only been an inspiration, but has also provided me with important feedback during the creation of this book.

I thank the Otterpine team for elevating this book to the next level, with a special thanks to Amy Reed, who has done a terrific job and shown her passion for the subject while editing this book.

I thank my friends for sharing with me the good and the bad moments.

Finally, I repeat words here that I have used in other publications of mine: I thank every scientist who has helped and continues to help to build science as we know it—all the people whose research is cited in this book, and distant scientists who contributed to the knowledge that humankind shares today. We are currently flirting with old ideas that endanger science, knowledge, and most importantly, the humble acceptance of our ignorance. May this book be another brick in the wall of our collective knowledge and another brick of resistance to such ideas.

NOTES

CHAPTER 1

1. "CLIMATE Action 30: 2023's Top Global Leaders Working Toward Solutions," Business Insider, November 27, 2023, https://www.businessinsider.com /climate-action-30-global-leaders-climate-solutions-2023-11.

2. Derek Muller, "Is Most Published Research Wrong?" posted August 11, 2016 by Veritasium YouTube channel, https://youtu.be/42QuXLucH3Q?si=GYAuba -tFh5YfxVj.

3. Leonard Mlodinow, *Subliminal: How Your Unconscious Mind Rules Your Behavior* (Allen Lane, 2012).

4. Steven Pinker, *Enlightenment Now: The Case for Reason, Science, Humanism, and Progress* (Viking, 2018).

5. C. Merchant, *The Death of Nature: Women, Ecology, and the Scientific Revolution* (HarperCollins Religious US, 1980).

6. Naomi Klein, *This Changes Everything: Capitalism vs. the Climate* (Simon & Schuster, 2015).

7. Yuval Noah Harari, "In the Battle Against Coronavirus, Humanity Lacks Leadership," *Time,* March 15, 2020, https://time.com/5803225/yuval-noah -harari-coronavirus-humanity-leadership/.

CHAPTER 2

8. Carlos Galindo Leal and Ibsen de Gusmão Câmara, eds., *The Atlantic Forest of South America: Biodiversity Status, Threats, and Outlook,* Center for Applied Biodiversity Science (Island Press, 2003).

9. "Mata Atlântica Biosphere Reserve, Brazil," United Nations Educational, Scientific and Cultural Organization (UNESCO), archived June 14, 2021, https:// web.archive.org/web/20210510110732/https://en.unesco.org/biosphere/lac /mata-atlantica.

10. "How Has Growth Changed Over Time?" Bank of England, accessed December 9, 2021, https://www.bankofengland.co.uk/knowledgebank/how-has-growth-changed-over-time.

11. WCED, *Report of the World Commission on Environment and Development: Our Common Future* (World Commission on Environment and Development [WCED], 1987).

12. M. K. Tolba and A. K. Biswas, *Earth and Us: Population–Resource–Environment–Development* (Elsevier, 2013).

13. S. Dietz and E. Neumayer, "Weak and Strong Sustainability in the SEEA: Concepts and Measurement," *Ecol. Econ.* 61, no. 4 (2007): 617–26.

14. E. Neumayer, *Weak Versus Strong Sustainability: Exploring the Limits of Two Opposing Paradigms,* 4th ed. (Edward Elgar Pub, 2013).

15. F. Ang and S. Van Passel, "Beyond the Environmentalist's Paradox and the Debate on Weak versus Strong Sustainability," *BioScience* 62, no. 3 (2012): 251–59.

16. S. Dietz and E. Neumayer, "Weak and Strong Sustainability in the SEEA: Concepts and Measurement," *Ecol. Econ.* 61, no. 4 (2007): 617–26.

17. Atila Iamarino, "Quanto vale o veneno de uma cobra jararaca?" ("How Much Is the Venom of a Jararaca Snake Worth?"), YouTube, posted October 25, 2019, accessed November 21, 2021, https://youtu.be/ymMVm7Rnaak.

18. F. Ang and S. Van Passel, "Beyond the Environmentalist's Paradox and the Debate on Weak versus Strong Sustainability," *BioScience* 62, no. 3 (2012): 251–59.

19. B. Bolin, *A History of the Science and Politics of Climate Change* (Cambridge University Press, 2008).

CHAPTER 4

20. L. K. Fazio et al., "Knowledge Does Not Protect Against Illusory Truth," *J. Exp. Psychol. Gen.* 144, no. 5 (2015): 993–1002.

21. D. Thompson, *Hit Makers: The Science of Popularity in an Age of Distraction* (Penguin Press, 2017).

22. P. C. Brown et al., *Make It Stick: The Science of Successful Learning* (Belknap Press: An Imprint of Harvard University Press, 2014).

23. N. Oreskes, "The Scientific Consensus on Climate Change," *Science* 306, no. 5702 (2004) 1686–1686.

24. J. Cook et al., "Consensus on Consensus: A Synthesis of Consensus Estimates on Human-Caused Global Warming," *Environ. Res. Lett.* 11, no. 4 (2016): 048002.

25. G. Marshall, *Don't Even Think About It: Why Our Brains Are Wired to Ignore Climate Change* (Bloomsbury Publishing, 2015).

26. M. E. Mann et al., "Northern Hemisphere Temperatures During the Past Millennium: Inferences, Uncertainties, and Limitations," *Geophys. Res. Lett.* 26, no. 6 (1999): 759–62.

27. PAGES2k Consortium, "A Global Multiproxy Database for Temperature Reconstructions of the Common Era," *Sci. Data* 4 (2017): 170088.

28. William McDonough, "Cradle to Cradle Design," TED Talks, Monterey, California, posted May 17, 2007, https://www.youtube.com/watch?v=IoRjz8iTVoo.

29. Hannah Ritchie and Max Roser, "Greenhouse Gas Emissions," Our World in Data, 2020, https://ourworldindata.org/greenhouse-gas-emissions.

30. IPCC, *Climate Change 2007: Synthesis Report. Contribution of Working Groups I, II and III to the Fourth Assessment Report of the Intergovernmental Panel on Climate Change*, Core Writing Team, R. K. Pachauri and A. Reisinger, eds. (Geneva, Switzerland: IPCC, 2007).

31. IPCC, *Climate Change 2007: Synthesis Report. Contribution of Working Groups I, II and III to the Fourth Assessment Report of the Intergovernmental Panel on Climate Change*, Core Writing Team, R. K. Pachauri and A. Reisinger, eds. (Geneva, Switzerland: IPCC, 2007).

32. E. G. Nisbet et al., "Very Strong Atmospheric Methane Growth in the 4 Years 2014–2017: Implications for the Paris Agreement," *Glob. Biogeochem. Cycles* 33, no. 3 (2019): 318–42.

33. E. G. Nisbet et al., "Very Strong Atmospheric Methane Growth in the 4 Years 2014–2017: Implications for the Paris Agreement," *Glob. Biogeochem. Cycles* 33, no. 3 (2019): 318–42.

34. V. Smil, *Energy: A Beginner's Guide* (Oneworld Publications, 2017).

35. M. A. K. Khalil and M. J. Shearer, "Sources of Methane: An Overview," in *Atmospheric Methane: Its Role in the Global Environment*, ed. M. A. K. Khalil (Springer, 2000): 98–111, https://doi:10.1007/978-3-662-04145-1_7.

36. R. B. Jackson et al., "Increasing Anthropogenic Methane Emissions Arise Equally from Agricultural and Fossil Fuel Sources," *Environ. Res. Lett.* 15, no. 7 (2020): 071002.

37. P. Bousquet et al., "Contribution of Anthropogenic and Natural Sources to Atmospheric Methane Variability," *Nature* 443 (2006): 439–43.

38. M. A. K. Khalil and M. J. Shearer, "Sources of Methane: An Overview," in *Atmospheric Methane: Its Role in the Global Environment*, ed. M. A. K. Khalil (Springer, 2000): 98–111, doi:10.1007/978-3-662-04145-1_7.

39. IPCC, ed. T.F. Stocker et al. *Climate Change 2013: The Physical Science Basis: Working Group I Contribution to the Fifth Assessment Report of the Intergovernmental Panel on Climate Change* (Cambridge University Press, 2013).

40. R. Guicherit and M. Roemer, "Tropospheric Ozone Trends," *Chemosphere - Glob. Change Sci.* 2, no. 2 (2000): 167–83.

41. EAA, *Tropospheric Ozone - Europe's Environment: The Dobris Assessment (Chapter 32)*, (European Environment Agency [EEA], 1994).

42. "Ground-level Ozone Basics," Environmental Protection Agency of the United States of America (USA EPA), last modified March 11, 2015, accessed June 4, 2021, https://www.epa.gov/ground-level-ozone-pollution/ground-level-ozone-basics.

43. IPCC, *Climate Change 2013: The Physical Science Basis: Working Group I Contribution to the Fifth Assessment Report of the Intergovernmental Panel on Climate Change*, ed. T.F. Stocker et al. (Cambridge University Press, 2013).

44. Royal Society (Great Britain) and D. Fowler, *Ground-Level Ozone in the 21st Century: Future Trends, Impacts and Policy Implications* (The Royal Society, 2008).

45. S. R. Pauleta et al., "Source and Reduction of Nitrous Oxide," *Coord. Chem. Rev.* 387 (2019): 436–49.

46. H. Tian et al., "A Comprehensive Quantification of Global Nitrous Oxide Sources and Sinks," *Nature* 586, 7828 (2020): 248–56.

47. S. R. Pauleta et al., "Source and Reduction of Nitrous Oxide," *Coord. Chem. Rev.* 387 (2019): 436–49.

48. H. Tian et al., "A Comprehensive Quantification of Global Nitrous Oxide Sources and Sinks," *Nature* 586, 7828 (2020): 248–56.

49. M. K. Firestone and E. A. Davidson, "Microbiological Basis of NO and N_2O Production and Consumption in Soil," *Exch. Trace Gases Terr. Ecosyst. Atmosphere* 47 (1989): 7–21.

50. H. Tian et al., "A Comprehensive Quantification of Global Nitrous Oxide Sources and Sinks," *Nature* 586 , 7828 (2020): 248–56.

51. "It's Water Vapor, Not the CO_2," American Chemical Society (ACS), accessed December 19, 2020, https://web.archive.org/web/20201209092627/ https://www.acs.org/content/acs/en/climatescience/climatesciencenarratives/its-water -vapor-not-the-co2.html.

52. "The NOAA Annual Greenhouse Gas Index (AGGI)," National Oceanic and Atmospheric Administration (NOAA), Global Monitoring Laboratory, US Department of Commerce, 2021, accessed June 26, 2021, https://gml.noaa .gov/aggi/aggi.html.

53. H. Ritchie, "Sector by Sector: Where Do Global Greenhouse Gas Emissions Come From?" Our World in Data, 2020, accessed March 27, 2021, https://ourworldindata.org/ghg-emissions-by-sector.

54. H. Ritchie and P. Rosado, "Electricity Mix," Our World in Data, 2020, accessed March 28, 2021, https://ourworldindata.org/electricity-mix.

55. A. Finkel, "Getting to Zero: Australia's Energy Transition," *Quarterly Essay* 81 (2021).

56. E. S. Cassidy et al., "Redefining Agricultural Yields: From Tonnes to People Nourished per Hectare," *Environ. Res. Lett.* 8, no. 3, (2013): 034015.

57. U.S. Energy Information Administration, "Chapter 8: Transportation Sector Energy Consumption" in *International Energy Outlook 2016 with Projections to 2040,* J. Conti et al. (U.S. Energy Information Administration, 2016), https://www.eia.gov/outlooks/archive/ieo16/pdf/0484(2016).pdf.

58. T. Fruergaard et al., "Energy Use and Recovery in Waste Management and Implications for Accounting of Greenhouse Gases and Global Warming Contributions," *Waste Manag. Res.* 27 no. 8 (2009): 724–37.

59. IPCC, "Chapter 3: Mobile Combustion" in *2006 IPCC Guidelines for National Greenhouse Gas Inventories,* ed. S. Eggleston et al., vol. 2 (IGES, 2006), https://www.ipcc-nggip.iges.or.jp/public/2006gl/pdf/2_Volume2/V2_3_Ch3_Mobile_Combustion.pdf.

60. H. Ritchie, "Sector by Sector: Where Do Global Greenhouse Gas Emissions Come From?" Our World in Data, 2020, accessed March 27, 2021, https://ourworldindata.org/ghg-emissions-by-sector.

61. B. Graver et al., CO2 Emissions from Commercial Aviation (ICCT20, 2019), https://theicct.org/publication/co2-emissions-from-commercial-aviation-2018/.

CHAPTER 5

62. S. Malavin et al., "Frozen Zoo: A Collection of Permafrost Samples Containing Viable Protists and Their Viruses," *Biodivers. Data J.* 8 (2020): e51586.

63. IPCC, *Climate Change 2014: Synthesis Report. Contribution of Working Groups I, II and III to the Fifth Assessment Report of the Intergovernmental Panel on Climate Change,* Core Writing Team, R. K. Pachauri, and L. A. Meyer, eds. (Geneva, Switzerland: IPCC, 2014).

64. R. N. Gibson et al., *Oceanography and Marine Biology: An Annual Review* (CRC Press, 2011).

65. P. Bhadury, "Effects of Ocean Acidification on Marine Invertebrates—a Review," *IJGMS* 44, no. 4 (April 2015): 1–11.

66. D. Allemand and D. Osborn, "Ocean Acidification Impacts on Coral Reefs: From Sciences to Solutions," *Reg. Stud. Mar. Sci.* 28, (2019): 100558.

67. IPCC, *Climate Change 2014: Synthesis Report. Contribution of Working Groups I, II and III to the Fifth Assessment Report of the Intergovernmental Panel on Climate Change*, Core Writing Team, R. K. Pachauri, and L. A. Meyer, eds. (Geneva, Switzerland: IPCC, 2014).

68. S. I. Seneviratne et al., "Chapter 3 - Changes in Climate Extremes and Their Impacts on the Natural Physical Environment," in *Managing the Risks of Extreme Events and Disasters to Advance Climate Change Adaptation: Special Report of the Intergovernmental Panel on Climate Change*, ed. C. B. Field et al. (Cambridge University Press, 2012): 109–230, https://doi:10.1017/CBO9781139177245.006.

69. S. I. Seneviratne et al., "Chapter 3 - Changes in Climate Extremes and Their Impacts on the Natural Physical Environment," in *Managing the Risks of Extreme Events and Disasters to Advance Climate Change Adaptation: Special Report of the Intergovernmental Panel on Climate Change*, ed. C. B. Field et al. (Cambridge University Press, 2012): 109–230, https://doi:10.1017/CBO9781139177245.006.

70. V. Weilnhammer et al., "Extreme Weather Events in Europe and Their Health Consequences – A Systematic Review," *Int. J. Hyg. Environ. Health* 233, 113688 (April 2021): 113688.

71. J.-M. Robine et al., "Death Toll Exceeded 70,000 in Europe During the Summer of 2003," *C. R. Biol.* 331, no. 2 (2008): 171–78.

72. M. M. Boer et al., "Unprecedented Burn Area of Australian Mega Forest Fires," *Nat. Clim. Change* 10 (February, 2020): 171–72.

73. G. J. van Oldenborgh et al., "Attribution of the Australian Bushfire Risk to Anthropogenic Climate Change," *Nat. Hazards Earth Syst. Sci.* 21, no. 3 (2021): 941–60.

74. S. C. Lewis et al., "Deconstructing Factors Contributing to the 2018 Fire Weather in Queensland, Australia," *Bull. Am. Meteorol. Soc.* 101, no. 1 (2020): S115–22.

75. S. I. Seneviratne et al., "Chapter 3 - Changes in Climate Extremes and Their Impacts on the Natural Physical Environment," in *Managing the Risks of Extreme Events and Disasters to Advance Climate Change Adaptation: Special Report of the Intergovernmental Panel on Climate Change*, ed. C. B. Field et al. (Cambridge University Press, 2012): 109–230, https://doi:10.1017/CBO9781139177245.006.

76. IPCC, *Climate Change 2014: Synthesis Report. Contribution of Working Groups I, II and III to the Fifth Assessment Report of the Intergovernmental Panel on Climate Change*, Core Writing Team, R. K. Pachauri, and L. A. Meyer, eds. (Geneva, Switzerland: IPCC, 2014).

77. Michael E. Mann and Tom Toles, *The Madhouse Effect: How Climate Change Denial Is Threatening Our Planet, Destroying Our Politics, and Driving Us Crazy* (Columbia University Press, 2016).

CHAPTER 6

78. ISO, *ISO 14040:2006. Environmental Management—Life Cycle Assessment—Principles and Framework, Eur. Comm. Stand.* (International Organization for Standardization, 2006).

79. M. P. Cenci et al., "Eco-Friendly Electronics—A Comprehensive Review," *Adv. Mater. Technol.* 7, no. 2, (2021): 2001263.

80. "Air Pollution," World Health Organization (WHO), 2021, accessed April 3, 2021, https://www.who.int/health-topics/air-pollution.

81. J. A. Bernstein et al., "Health Effects of Air Pollution," *J. Allergy Clin. Immunol.* 114 no. 5 (2004): 1116–23.

82. "Particulate Matter (PM10 and PM2.5)," NSW (New South Wales) Health, NSW.gov, last modified November 25, 2020, accessed April 5, 2021, https://www.health.nsw.gov.au/environment/air/Pages/particulate-matter.aspx.

83. M. Kampa and E. Castanas, "Human Health Effects of Air Pollution," *Environ. Pollut.* 151 no. 2 (2008): 362–67.

84. "Household Air Pollution," WHO (World Health Organization), last modified 2018, accessed April 5, 2021, https://www.who.int/news-room/fact-sheets/detail/household-air-pollution-and-health.

85. V. Smil, *Energy: A Beginner's Guide* (Oneworld Publications, 2017).

86. V. Smil, *Energy: A Beginner's Guide* (Oneworld Publications, 2017).

87. V. Smil, *Energy: A Beginner's Guide* (Oneworld Publications, 2017).

88. T. Fruergaard et al., "Energy Use and Recovery in Waste Management and Implications for Accounting of Greenhouse Gases and Global Warming Contributions," *Waste Manag. Res.* 27, no. 8 (2009): 724–37.

89. IPCC, "Chapter 3: Mobile Combustion" in *2006 IPCC Guidelines for National Greenhouse Gas Inventories*, ed. H. Eggleson et al., vol. 2 (IGES, 2006), https://www.ipcc-nggip.iges.or.jp/public/2006gl/pdf/2_Volume2/V2_3_Ch3_Mobile_Combustion.pdf.

90. V. Smil, *Energy: A Beginner's Guide* (Oneworld Publications, 2017).

91. K. Skalska et al., "Trends in NO_X Abatement: A Review," *Sci. Total Environ.* 408, no. 19 (2010): 3976–89.

92. P. R. Bartrop and P. S. Totten, *Dictionary of Genocide* (Greenwood, 2007).

93. "Household Air Pollution," WHO (World Health Organization), last modified 2018, accessed April 5, 2021, https://www.who.int/news-room/fact-sheets/detail/household-air-pollution-and-health.

94. P. D. Noyes et al., "The Toxicology of Climate Change: Environmental Contaminants in a Warming World," *Environ. Int.* 35, no. 6 (2009): 971–86.

95. Y. Zhang et al., "Tracing Nitrate Pollution Sources and Transformation in Surface- and Ground-waters Using Environmental Isotopes," *Sci. Total Environ.* 490 (August 2014): 213–22.

96. C. Schaum, *Phosphorus: Polluter and Resource of the Future: Motivations, Technologies and Assessment of the Elimination and Recovery of Phosphorus from Wastewater* (IWA Publishing, 2018).

97. M. Goedkoop et al., *ReCiPe 2008: A Life Cycle Impact Assessment Method Which Comprises Harmonised Category Indicators at the Midpoint and the Endpoint Level* (The Hague, Ministry of VROM, 2009).

98. F. Verones et al., *LC-IMPACT Version 1.0: A Spatially Differentiated Life Cycle Impact Assessment Approach* (LC-IMPACT, 2020), https://www.lc-impact.eu/about.html.

99. E. L. Zvereva et al., "Changes in Species Richness of Vascular Plants Under the Impact of Air Pollution: A Global Perspective," *Glob. Ecol. Biogeogr.* 17, no. 3 (2008): 305–19.

100. Neville Fletcher, "Earth's Sunscreen, the Ozone Layer," Australian Academy of Science, last modified 2015, accessed February 12, 2021, https://www.science.org.au/curious/earth-environment/earths-sunscreen-ozone-layer.

101. A. M. Jubb et al., "An Atmospheric Photochemical Source of the Persistent Greenhouse Gas CF_4," *Geophys. Res. Lett.* 42, no. 21 (2015): 9505–11.

102. "Some Chemicals Less Damaging to Ozone Can Degrade to Long-lived Greenhouse Gas," American Geophysical Union, ScienceDaily, November 3, 2015, https://www.sciencedaily.com/releases/2015/11/151103140111.htm.

103. M. Goedkoop et al., *ReCiPe 2008: A Life Cycle Impact Assessment Method Which Comprises Harmonised Category Indicators at the Midpoint and the Endpoint Level* (The Hague, Ministry of VROM, 2009).

104. Neville Fletcher, "Earth's Sunscreen, the Ozone Layer," Australian Academy of Science, last modified 2015, accessed February 12, 2021, https://www.science .org.au/curious/earth-environment/earths-sunscreen-ozone-layer.

105. Mario Molina, "The Montreal Protocol: Triumph by Treaty," UNEP, November 20, 2017, http://www.unenvironment.org/news-and-stories/story /montreal-protocol-triumph-treaty.

106. M. Goedkoop et al., *ReCiPe 2008: A Life Cycle Impact Assessment Method Which Comprises Harmonised Category Indicators at the Midpoint and the Endpoint Level* (The Hague, Ministry of VROM, 2009).

107. M. F. Chislock et al., "Eutrophication: Causes, Consequences, and Controls in Aquatic Ecosystems," *Nat. Educ. Knowl.* 4, no. 4 (2013): 10.

108. M. Goedkoop et al., *ReCiPe 2008: A Life Cycle Impact Assessment Method Which Comprises Harmonised Category Indicators at the Midpoint and the Endpoint Level* (The Hague, Ministry of VROM, 2009).

109. M. F. Chislock et al., "Eutrophication: Causes, Consequences, and Controls in Aquatic Ecosystems," *Nat. Educ. Knowl.* 4, no. 4 (2013): 10.

110. "Colwyn Bay Man Died After 200-Coffee Caffeine Overdose," BBC News, March 1, 2022, https://www.bbc.com/news/uk-wales-60570470.

111. M. Goedkoop et al., *ReCiPe 2008: A Life Cycle Impact Assessment Method Which Comprises Harmonised Category Indicators at the Midpoint and the Endpoint Level* (The Hague, Ministry of VROM, 2009).

112. A. D. Henderson et al., "USEtox Fate and Ecotoxicity Factors for Comparative Assessment of Toxic Emissions in Life Cycle Analysis: Sensitivity to Key Chemical Properties," *Int. J. Life Cycle Assess.* 16, no. 8 (2011) 701–709.

113. Michael Z. Hauschild and Mark A. J. Hujibregts, eds, *Life Cycle Impact Assessment*, (Springer Netherlands, 2015), https://doi:10.1007/978-94-017-9744-3.

114. A. D. Henderson et al., "USEtox Fate and Ecotoxicity Factors for Comparative Assessment of Toxic Emissions in Life Cycle Analysis: Sensitivity to Key Chemical Properties," *Int. J. Life Cycle Assess.* 16, iss. 8 (2011).

115. Michael Z. Hauschild and Mark A. J. Hujibregts, eds, *Life Cycle Impact Assessment*, (Springer Netherlands, 2015), https://doi:10.1007/978-94-017-9744-3.

116. E. Mann et al., "Mercury Fate in Ageing and Melting Snow: Development and Testing of a Controlled Laboratory System," *J. Environ. Monit.* 13, no. 10 (2011): 2695–2702.

117. P. D. Noyes et al., "The Toxicology of Climate Change: Environmental Contaminants in a Warming World," *Environ. Int.* 35, no. 6 (2009): 971–86.

118. P. D. Noyes et al., "The Toxicology of Climate Change: Environmental Contaminants in a Warming World," *Environ. Int.* 35, iss. 6 (2009): 971–86.

119. Christina M. Kennedy et al., "Managing the Middle: A Shift in Conservation Priorities Based on the Global Human Modification Gradient," Global Change Biology 25, no. 3 (2019): 811–26, https://doi.org/10.1111/gcb.14549.

120. Mark A. J. Hujibregts et al., "ReCiPe2016: A Harmonised Life Cycle Impact Assessment Method at Midpoint and Endpoint Level," *The International* Journal of Life Cycle Assessment 22, no. 2 (2017): 138–147, accessed September 8, 2024, DOI: https://doi.org/10.1007/s11367-016-1246-y.

121. F. S. Chapin III et al., "Consequences of Changing Biodiversity," *Nature* 405 (May, 2000): 234–42.

122. R. H. Cowie et al., "The Sixth Mass Extinction: Fact, Fiction or Speculation?" *Biol. Rev.* 97, no. 2 (2022): 640–63.

123. N. L. Waller et al., "The Bramble Cay melomys Melomys rubicola (Rodentia: Muridae): A First Mammalian Extinction Caused by Human-Induced Climate Change?" *Wildl. Res.* 44, no. 1 (2017): 9–21.

124. J. A. Pounds, "Climate and Amphibian Declines," *Nature* 410 (April, 2001): 639–40.

125. R. H. Cowie et al., "The Sixth Mass Extinction: Fact, Fiction or Speculation?" *Biol. Rev.* 97, no. 2 (2022): 640–63.

126. "Summary Statistics," The IUCN Red List of Threatened Species, IUCN, last modified 2021, accessed March 4, 2021, https://www.iucnredlist.org/resources /summary-statistics.

127. WWF, *Living Planet Report 2020: Bending the Curve of Biodiversity Loss*, ed. R.E.A. Almond et al. (Gland, Switzerland: WWF, 2020)

128. J. W. Bull and N. Strange, "The Global Extent of Biodiversity Offset Implementation Under No Net Loss Policies," *Nat. Sustain.* 1 (November, 2018): 790–98.

129. C. H. Trisos et al., "The Projected Timing of Abrupt Ecological Disruption from Climate Change," *Nature* 580 (April, 2020): 496–501.

130. J. B. Guinée and R. Heijungs, "A Proposal for the Definition of Resource Equivalency Factors for Use in Product Life-Cycle Assessment," *Environ. Toxicol. Chem.* 14, no. 5 (1995): 917–25.

131. Michael Z. Hauschild et al., eds, *Life Cycle Assessment: Theory and Practice* (Springer International Publishing, 2018), https://doi:10.1007/978-3-319-56475-3.

132. J. B. Guinée and R. Heijungs, "A Proposal for the Definition of Resource Equivalency Factors for Use in Product Life-Cycle Assessment," *Environ. Toxicol. Chem.* 14, iss. 5 (1995): 917–25.

133. L. V. Oers and J. Guinée, "The Abiotic Depletion Potential: Background, Updates, and Future," *Resources* 5, no. 1 (2016): 1–12.

134. T. Sowell, *Basic Economics* (Basic Books, 2014).

135. J. B. Guinée and R. Heijungs, "A Proposal for the Definition of Resource Equivalency Factors for Use in Product Life-Cycle Assessment," *Environ. Toxicol. Chem.* 14, no. 5 (1995): 917–25.

136. UNEP, *Sand and Sustainability: Finding New Solutions for Environmental Governance of Global Sand Resources* (United Nations Environment Programme [UNEP], 2019).

137. L. V. Oers and J. Guinée, "The Abiotic Depletion Potential: Background, Updates, and Future," *Resources* 5, no. 1 (2016): 1–12.

138. USGS, "Principles of a Resource/reserve Classification for Minerals," *US Geol. Surv. Circ.* 831 (1980): 5.

139. D. Brown, *Inferno* (Anchor Books, 2014).

140. H. Rosling et al., *Factfulness: Ten Reasons We're Wrong About the World—and Why Things Are Better Than You Think* (Flatiron Books, 2018).

141. UN, "How Certain Are the United Nations Global Population Projections?" *Population Facts* No. 2019/6, Population Division, United Nations Department of Economic and Social Affairs, December 2019, accessed May 16, 2022, https://www.un.org/en/development/desa/population/publications/pdf/popfacts/PopFacts_2019-6.pdf.

142. Hans Rosling, "Religion and Babies," TED Talks, posted May 22, 2012, https://www.youtube.com/watch?v=ezVk1ahRF78

143. Free data from World Bank via Gapminder.org, dataset accessed October 12, 2024, from https://www.gapminder.org/data/. Free to use! CC-BY Gapminder.org.

144. S. E. Vollset et al., "Fertility, Mortality, Migration, and Population Scenarios for 195 Countries and Territories from 2017 to 2100: A Forecasting Analysis for the Global Burden of Disease Study," *The Lancet* 396, no. 10258 (2020): P 1285–1306.

145. Håvard Egge, "Antidepressants and Painkillers Found in Crustaceans on Svalbard" ("Fant antidepressiva og smertestillende i krepsdyr på Svalbard"), SINTEF, posted January 20, 2021, https://www.sintef.no/siste-nytt/2021/fant-antidepressiva-og-smertestillende-i-krepsdyr-pa-svalbard/.

146. IDMC, 2016 Global Report on Internal Displacement (IDMC, May 2016), https://api.internal-displacement.org/sites/default/files/publications /documents/2016-global-report-internal-displacement-IDMC.pdf.

147. IPCC, *Climate Change 2023: Synthesis Report. Contribution of Working Groups I, II and III to the Sixth Assessment Report of the Intergovernmental Panel on Climate Change*, Core Writing Team, H. Lee, and J. Romero, eds. (Geneva, Switzerland: IPCC, 2023): 35–115, https://doi:10.59327/IPCC/ AR6-9789291691647.

148. "Fires," NASA Earth Observatory website, https://earthobservatory.nasa.gov /topic/fires.

CHAPTER 7

149. D. Kahneman, *Thinking, Fast and Slow* (Farrar, Straus and Giroux, 2013).

150. G. Marshall, *Don't Even Think About It: Why Our Brains Are Wired to Ignore Climate Change* (Bloomsbury, 2015).

151. G. Marshall, *Don't Even Think About It: Why Our Brains Are Wired to Ignore Climate Change* (Bloomsbury Publishing, 2015).

152. M. E. Mann, *The New Climate War: The Fight to Take Back Our Planet* (Scribe Publications, 2021).

153. K. Niinimäki et al., "The Environmental Price of Fast Fashion," *Nat Rev Earth Environ* 1 (April, 2020): 189–200, https://www.nature.com/articles /s43017-020-0039-9.

154. S. Pape, *The Barefoot Investor 2020 Update: The Only Money Guide You'll Ever Need* (Wiley, 2020).

155. F.-R. Yang et al., "Effect of Facebook Social Comparison on Well-being: A Meta-Analysis," *J. Internet Technol.* 20, no. 6 (2019): 1829–36.

156. J. Fardouly et al., "Social Comparisons on Social Media: The Impact of Facebook on Young Women's Body Image Concerns and Mood," *Body Image* 13 (March, 2015): 38–45.

157. H. Ritchie, "Sector by Sector: Where Do Global Greenhouse Gas Emissions Come From?" Our World in Data, last modified 2020, accessed March 27, 2021, https: //ourworldindata.org/ghg-emissions-by-sector.

158. B. Graver et al., CO2 Emissions from Commercial Aviation (ICCT20, 2019), https://theicct.org/publication/co2-emissions-from-commercial-aviation -2018/.

159. T. Bralower and D. Bice, "Distribution of Water on the Earth's Surface," Earth in the Future, PSU, last modified 2012, accessed September 14, 2021, https://www.e-education.psu.edu/earth103/node/701.

160. Water Science School, "Where Is Earth's Water?" USGS, posted June 6, 2018, accessed September 14, 2021, https://www.usgs.gov/special-topic/water-science-school/science/where-earths-water?qt-science_center_objects=0#qt-science_center_objects.

161. Institute of Medicine, *Dietary Reference Intakes for Water, Potassium, Sodium, Chloride, and Sulfate* (The National Academies Press, 2005), https://doi.org/10.17226/10925.

162. H. Ritchie and M. Roser, "Water Use and Stress: Share of Freshwater Withdrawals Used in Agriculture," Our World in Data, last updated 2017, accessed August 27, 2021, https://ourworldindata.org/water-use-stress.

163. P. H. Gleick et al., *The World's Water Volume 7: The Biennial Report on Freshwater Resources* (Island Press, 2012): 237.

164. ANA, *Manual de usos consuntivos da água no Brasil* (*Manual of Consumptive Uses of Water in Brazil*) (Agencia Nacional de Aguas [ANA], 2019). Document in Portuguese; free translation: ANA - National Water Agency.

165. ANA, *Manual de usos consuntivos da água no Brasil* (*Manual of Consumptive Uses of Water in Brazil*) (Agencia Nacional de Aguas [ANA], 2019). Document in Portuguese; free translation: ANA - National Water Agency.

166. S. Karayi, *PC Energy Report 2009:* United States, United Kingdom, Germany, (1E, 2009): 20, accessed September 14, 2021, https://www.twosides.info/wp-content/uploads/2018/05/The_Power_to_Save_Money.pdf.

167. J. Abraham et al., "Green Fog Computing: A Review on the Basis of Latency, Energy, and e-Waste," *Int. J. Adv. Sci. Technol.* 29, no. 3 (2020): 5617–25.

168. S. Karayi, *PC Energy Report 2009: United States, United Kingdom, Germany,* (1E, 2009): 20, accessed September 14, 2021, https://www.twosides.info/wp-content/uploads/2018/05/The_Power_to_Save_Money.pdf.

169. "How Many Computers Are There in the World?" SCMO, posted August 9, 2019, https://www.scmo.net/faq/2019/8/9/how-many-compaters-is-there-in-the-world.

170. NTIA, "Chapter 6: The Digital Workplace," in *A Nation Online: How Americans Are Expanding Their Use of the Internet* (National Telecommunications and Information Administration [NTIA], 2001), https://web.archive.org/web/20220928130654/https://www.ntia.doc.gov/legacy/ntiahome/dn/html/Chapter6.htm.

171. "Country Comparisons—Labor Force," The World Factbook, Central Intelligence Agency (CIA) website, last updated 2020, https://www.cia.gov/the-world -factbook/field/labor-force/country-comparison.

172. M. P. Cenci et al., "Eco-Friendly Electronics—A Comprehensive Review," *Adv. Mater. Technol.* 7, no. 2, (2022): 2001263.

173. H. Rosling et al., *Factfulness: Ten Reasons We're Wrong About the World—and Why Things Are Better Than You Think* (Flatiron Books, 2018).

174. J. Poore and T. Nemecek, "Reducing Food's Environmental Impacts Through Producers and Consumers," *Science* 360 (2018): 987–92.

175. J. Poore and T. Nemecek, "Reducing Food's Environmental Impacts Through Producers and Consumers," *Science* 360, no. 6392 (2018): 987–92.

176. "Land Use in Agriculture by the Numbers," Food and Agriculture Organization of the United Nations (FAO), last modified May 7, 2020, accessed March 30, 2022, https://www.fao.org/sustainability/news/detail/en/c/1274219/.

177. J. Poore and T. Nemecek, "Reducing Food's Environmental Impacts Through Producers and Consumers," *Science* 360 (2018): 987–92.

178. "Land Use in Agriculture by the Numbers," Food and Agriculture Organization of the United Nations (FAO), May 7, 2020, accessed March 30, 2022, https: //www.fao.org/sustainability/news/detail/en/c/1274219/.

179. J. G. Fadel, "Quantitative Analyses of Selected Plant By-Product Feedstuffs, a Global Perspective," *Anim. Feed Sci. Technol.* 79, no. 4 (1999): 255–68.

180. M. Clark and D. Tilman, "Comparative Analysis of Environmental Impacts of Agricultural Production Systems, Agricultural Input Efficiency, and Food Choice," *Environ. Res. Lett.* 12, no. 6 (2017): 064016.

181. H. Ritchie, "Is Organic Really Better for the Environment Than Conventional Agriculture?" Our World in Data, last modified 2017, accessed July 16, 2021, https://ourworldindata.org/is-organic-agriculture-better-for-the-environment.

182. A. Sulleyman, "Pro Donald Trump 4chan Group Reporting Illegal US-Mexico Border Crossings 'Using Network Webcams,'" Independent, published March 21, 2017, accessed September 21, 2021, https://www.independent.co.uk /life-style/gadgets-and-tech/news/pro-donald-trump-us-mexico-border-4chan -group-illegal-crossing-network-webcams-blueservo-a7641031.html.

183. "A Amazônia em números" ("The Amazon in Numbers"), Imazon, published June 23, 2009, https://imazon.org.br/imprensa/a-amazonia-em-numeros/.

184. D. Nepstad et al., "Slowing Amazon Deforestation Through Public Policy and Interventions in Beef and Soy Supply Chains," *Science* 344 no. 6188 (2014): 1118–23.

185. C. H. L. Silva Jr. et al., "The Brazilian Amazon Deforestation Rate in 2020 Is the Greatest of the Decade," *Nat. Ecol. Evol.* 5 (December, 2021): 144–45.

186. M. M. Vale et al., "The COVID-19 Pandemic as an Opportunity to Weaken Environmental Protection in Brazil," *Biol. Conserv.* 255, (2021): 108994.

187. M. M. Vale et al., "The COVID-19 Pandemic as an Opportunity to Weaken Environmental Protection in Brazil," *Biol. Conserv.* 255, (2021): 108994.

188. M. C. Marcello and J. Spring, "Brazil Environment Minister Quits; Faces Illegal Logging Probe," *Reuters,* June 24, 2021, https://www.reuters.com/world/americas/brazil-environment-minister-salles-resigns-amid-illegal-logging-probe-2021-06-23/.

189. M. E. Mann, *The New Climate War: The Fight to Take Back Our Planet* (Scribe Publications, 2021).

190. N. Klein, *This Changes Everything: Capitalism vs. the Climate* (Simon & Schuster, 2015).

191. UNSW Media, "UNSW Steps Up Action on Climate Change," UNSW Newsroom, published March 5, 2020, accessed May 3, 2021, https://newsroom.unsw.edu.au/news/general/unsw-steps-action-climate-change.

192. N. Klein, *This Changes Everything: Capitalism vs. the Climate* (Simon & Schuster, 2015).

193. N. Chomsky, *Understanding Power: The Indispensable Chomsky* (The New Press, 2002).

194. "Screening of Websites for 'Greenwashing': Half of Green Claims Lack Evidence," European Commission, press release, published January 27, 2021, accessed May 15, 2022, https://ec.europa.eu/commission/presscorner/detail/en/ip_21_269.

195. "Global Sweep Finds 40% of Firms' Green Claims Could Be Misleading," Competition and Markets Authority, GOV.UK, press release, published January 28, 2021, accessed May 15, 2022, https://www.gov.uk/government/news/global-sweep-finds-40-of-firms-green-claims-could-be-misleading.

196. E. Chenoweth et al., *Why Civil Resistance Works: The Strategic Logic of Nonviolent Conflict* (Columbia University Press, 2011).

197. B. Gates, *How to Avoid a Climate Disaster: The Solutions We Have and the Breakthroughs We Need* (Random House Large Print, 2021).

198. P. Dias et al., "Comprehensive Recycling of Silicon Photovoltaic Modules Incorporating Organic Solvent Delamination – Technical, Environmental and Economic Analyses," *Resour. Conserv. Recycl.* 165, (February, 2021): 105241.

199. T. Sowell, *Basic Economics* (Basic Books, 2014).

200. Jennifer Uju Okonkwo, "Welfare Effects of Carbon Taxation on South African Households," *Energy Economics* 96, (April, 2021): 104903: 104903, https://www.sciencedirect.com/science/article/abs/pii/S0140988320302437.

201. Franziska Funke and Linus Mattauch, "Why Is Carbon Pricing in Some Countries More Successful Than in Others?" Our World in Data, published August 10, 2018, https://ourworldindata.org/carbon-pricing-popular.

202. Franziska Funke and Linus Mattauch, "Why Is Carbon Pricing in Some Countries More Successful Than in Others?" Our World in Data, published August 10, 2018, https://ourworldindata.org/carbon-pricing-popular.

203. Angela Köppl and Margit Schratzenstaller, "Carbon Taxation: A Review of the Empirical Literature," *Journal of Economic Surveys* 37, no. 4 (2023): 1353–88, https://onlinelibrary.wiley.com/doi/full/10.1111/joes.12531.

204. Jennifer Uju Okonkwo, "Welfare Effects of Carbon Taxation on South African Households," *En*ergy Economics 96, (April, 2021): 104903 , https://www.sciencedirect.com/science/article/abs/pii/S0140988320302437.

205. Jennifer Uju Okonkwo, "Welfare Effects of Carbon Taxation on South African Households," *Energy Eco*nomics 96, 104903 (2021), https://www.sciencedirect.com/science/article/abs/pii/S0140988320302437.

206. Franziska Funke and Linus Mattauch, "Why Is Carbon Pricing in Some Countries More Successful Than in Others?" Our World in Data, published August 10, 2018, https://ourworldindata.org/carbon-pricing-popular.

207. M. Jaccard, *The Citizen's Guide to Climate Success: Overcoming Myths that Hinder Progress* (Cambridge University Press, 2020), https://doi:10.1017/9781108783453.

208. Joana Setzer and Lisa Benjamin, "Climate Change Litigation in the Global South: Filling in Gaps," *AJIL Unbound* 114 (February, 2020) 56–60, DOI: https://doi.org/10.1017/aju.2020.6.

209. DW Planet A, "Why Fossil Fuel Companies Should Be Lawyering Up," YouTube posted July 30, 2021, accessed on February 19, 2022, https://www.youtube.com/watch?v=yVYzHgHx8U4.

210. C. Ionescu et al., "The Historical Evolution of the Energy Efficient Buildings," *Renew. Sustain. Energy Rev.* 49 (September, 2015): 243–53.

211. The Economist Intelligence Unit (EIU), "An Eco-wakening: Measuring Global Awareness, Engagement and Action for Nature," Economist Impact, published September 29, 2021, accessed August 22, 2021, https://impact

.economist.com/sustainability/ecosystems-resources/an-eco-wakening
-measuring-global-awareness-engagement-and-action-for-nature.

212. N. Landreth et al., "Assessing the Effectiveness of Building Simulation to Regulate Residential Water Consumption and Greenhouse Gas Emissions in New South Wales, Australia," *12th Conference of International Building Performance Simulation Association* (2011).

213. Paris A. Fokaides, Kyriacos Polycarpou, and Soteris Kalogirou, "The Impact of the Implementation of the European Energy Performance of Buildings Directive on the European Building Stock: The Case of the Cyprus Land Development Corporation," Energy Policy 111 (2017), https://doi.org/10.1016/j.enpol.2017.09.009.

214. L. M. López-Ochoa et al., "Towards Nearly Zero-Energy Buildings in Mediterranean Countries: Fifteen Years of Implementing the Energy Performance of Buildings Directive in Spain (2006–2020)," *J. Build. Eng.* 44, (December, 2021): 102962.

215. "Energy Performance of Buildings Directive," European Commission (EC), published May 16, 2019, archived August 17, 2021, https://web.archive.org/web/20210814115502/https://ec.europa.eu/energy/topics/energy-efficiency/energy-efficient-buildings/energy-performance-buildings-directive_en#facts-and-figures.

216. Kirsten Dirksen, "Tokyo Office Grows Own Food in Vertical Farm," YouTube posted January 25, 2016, accessed August 10, 2021, https://youtu.be/qJMZRIRkZWs.

217. "Pasona Urban Farm," KONODESIGNS, accessed August 10, 2021, http://konodesigns.com/urban-farm/.

218. World Bank Open Data, "Urban Population (% of Total Population). United Nations Population Division. All Countries and Economies Most Recent Values (2020 & 2021)," dataset accessed August 19, 2021, https://data.worldbank.org.

219. UN, "The Speed of Urbanization Around the World," United Nations Department of Economic and Social Affairs (UN DESA), *Population Facts* 1, (December, 2018): 1–2.

CHAPTER 8

220. T. Sowell, *Basic Economics* (Basic Books, 2014).

221. N. Klein, *This Changes Everything: Capitalism vs. the Climate* (Simon & Schuster, 2015).

222. A. Zalik, "Resource Sterilization: Reserve Replacement, Financial Risk, and Environmental Review in Canada's Tar Sands," *Environ. Plan. Econ. Space* 47, no. 12 (2015): 2446–64.

223. W. S. Jevons, *The Coal Question; An Inquiry Concerning the Progress of the Nation, and the Probable Exhaustion of Our Coal-Mines* (Macmillan and Co., 1865).

224. P. Smith et al., "Biophysical and Economic Limits to Negative CO_2 Emissions," *Nat. Clim. Change* 6 (December, 2015): 42–50.

225. D. Woolf et al., "Sustainable Biochar to Mitigate Global Climate Change," *Nat. Commun.* 1, 56 (2010).

226. "Carbon Capture, Utilisation and Storage," International Energy Agency (IEA), last modified 2022, https://www.iea.org/fuels-and-technologies/carbon -capture-utilisation-and-storage.

227. IEA, "CCUS in Clean Energy Transitions," *Energy Technology Perspectives 2020: Special Report on Carbon Capture Utilisation and Storage* (International Energy Agency [IEA], 2020), https://www.iea.org/reports/ccus-in-clean -energy-transitions.

228. P. Smith et al., "Biophysical and Economic Limits to Negative CO_2 Emissions," *Nat. Clim. Change* 6 (December, 2016): 42–50.

229. S. H. Widder et al., *Sustainability Assessment of Coal-Fired Power Plants with Carbon Capture and Storage*, October 2011, U.S. Department of Energy, PPNNL-20933, https://doi:10.2172/1031992.

230. M. Broehm et al., "Techno-Economic Review of Direct Air Capture Systems for Large Scale Mitigation of Atmospheric CO2," posted September 26, 2015, https://doi:10.2139/ssrn.2665702.

231. M. Broehm et al., "Techno-Economic Review of Direct Air Capture Systems for Large Scale Mitigation of Atmospheric CO2," posted September 26, 2015, https://doi:10.2139/ssrn.2665702.

232. F. Swain, "The Device That Reverses CO2 Emissions," BBC, posted March 11, 2021, https://www.bbc.com/future/article/20210310-the-trillion-dollar -plan-to-capture-co2.

233. M. Broehm et al., "Techno-Economic Review of Direct Air Capture Systems for Large Scale Mitigation of Atmospheric CO2," posted September 26, 2015, https://doi:10.2139/ssrn.2665702.

234. G. Realmonte et al., "An Inter-Model Assessment of the Role of Direct Air Capture in Deep Mitigation Pathways," *Nat. Commun.* 10 (2019): 3277.

235. G. Realmonte et al., "An Inter-Model Assessment of the Role of Direct Air Capture in Deep Mitigation Pathways," *Nat. Commun.* 10, 3277 (2019).

236. G. Realmonte et al., "An Inter-Model Assessment of the Role of Direct Air Capture in Deep Mitigation Pathways," *Nat. Commun.* 10 (2019): 3277.

237. DW Planet A, "Made of Pollution: How CO2 Is Recycled to Make Your Things," YouTube posted October 15, 2021, https://youtu.be/9wESzQ0-ZjQ.

238. B. Parkin, "Swiss Pickles Set to Benefit From First Carbon Capture Plant," Bloomberg.com, published May 31, 2017, https://www.bloomberg.com/news/articles/2017-05-31/swiss-pickles-set-to-benefit-from-first-carbon-capture-plant.

239. P. Smith et al., "Biophysical and Economic Limits to Negative CO_2 Emissions," *Nat. Clim. Change* 6 (December, 2015): 42–50.

240. C. Zhou et al., "Impacts of a Large-Scale Reforestation Program on Carbon Storage Dynamics in Guangdong, China," *For. Ecol. Manag.* 255, no. 3–4 (2008): 847–54.

241. J. Fang et al., "Changes in Forest Biomass Carbon Storage in China Between 1949 and 1998," *Science* 292, no. 5525 (2001): 2320–22.

242. *Greenhouse Gas Removal* (Royal Society [Great Britain] & Royal Academy of Engineering [Great Britain], 2018) https://royalsociety.org/-/media/policy/projects/greenhouse-gas-removal/royal-society-greenhouse-gas-removal-report-2018.pdf.

243. P. Smith et al., "Biophysical and Economic Limits to Negative CO_2 Emissions," *Nat. Clim. Change* 6 (December, 2015): 42–50.

244. G. Bala et al., "Combined Climate and Carbon-Cycle Effects of Large-Scale Deforestation," *Proc. Natl. Acad. Sci.* 104, no. 16 (2007): 6550–55.

245. J.-F. Bastin et al., "The Global Tree Restoration Potential," *Science* 365, no. 6448 (2019): 76–79 https://doi.org/10.1126/science.aax0848.

246. Abhijit V. Banerjee and Esther Duflo, *Good Economics for Hard Times*, (PublicAffairs, 2019).

247. D. J. Beerling et al., "Farming with Crops and Rocks to Address Global Climate, Food and Soil Security," *Nat. Plants* 4 (February, 2018): 138–47.

248. J. Hartmann et al., "Global CO_2-Consumption by Chemical Weathering: What Is the Contribution of Highly Active Weathering Regions?" *Glob. Planet. Change* 69, no. 4 (2009): 185–94.

249. D. S. Goll et al., "Potential CO_2 Removal from Enhanced Weathering by Ecosystem Responses to Powdered Rock," *Nat. Geosci.* 14 (July, 2021): 545–49.

250. D. Beerling and S. Long, "Guest Post: How 'Enhanced Weathering' Could Slow Climate Change and Boost Crop Yields," Carbon Brief, posted February 19, 2018, accessed December 6, 2021, https://www.carbonbrief.org/guest-post -how-enhanced-weathering-could-slow-climate-change-and-boost-crop-yields.

251. L. L. Taylor et al., "Enhanced Weathering Strategies for Stabilizing Climate and Averting Ocean Acidification," *Nat. Clim. Change* 6 (December, 2016): 402–6.

252. P. Renforth, "The Negative Emission Potential of Alkaline Materials," *Nat. Commun.* 10, (2019): 1401.

253. D. S. Goll et al., "Potential CO_2 Removal from Enhanced Weathering by Ecosystem Responses to Powdered Rock," *Nat. Geosci.* 14 July, (2021): 545–49.

254. L. L. Taylor et al., "Enhanced Weathering Strategies for Stabilizing Climate and Averting Ocean Acidification," *Nat. Clim. Change* 6 (December, 2016): 402–6.

255. D. S. Goll et al., "Potential CO_2 Removal from Enhanced Weathering by Ecosystem Responses to Powdered Rock," *Nat. Geosci.* 14 (July, 2021): 545–49.

256. D. S. Goll et al., "Potential CO_2 Removal from Enhanced Weathering by Ecosystem Responses to Powdered Rock," *Nat. Geosci.* 14 (2021): 545–49.

257. S. D. Levitt and S. J. Dubner, *SuperFreakonomics: Global Cooling, Patriotic Prostitutes, and Why Suicide Bombers Should Buy Life Insurance* (William Morrow Paperbacks, 2011).

258. N. Klein, *This Changes Everything: Capitalism vs. the Climate* (Simon & Schuster, 2015).

259. A. Robock et al., "A Test for Geoengineering?" *Science* 327, no. 5965 (2010): 530–1.

260. A. Robock et al., "Regional Climate Responses to Geoengineering with Tropical and Arctic SO_2 Injections," *J. Geophys. Res. Atmospheres* 113, no. D16 (2008).

261. N. Klein, *This Changes Everything: Capitalism vs. the Climate* (Simon & Schuster, 2015).

262. J. M. Haywood et al., "Asymmetric Forcing from Stratospheric Aerosols Impacts Sahelian Rainfall," *Nat. Clim. Change* 3 March, (2013): 660–65.

263. A. Robock et al., "A Test for Geoengineering?" *Science* 327, (2010): 530–1.

264. A. Robock et al., "Regional Climate Responses to Geoengineering with Tropical and Arctic SO_2 Injections," *J. Geophys. Res. Atmospheres* 113, no. D16 (2008).

265. K. E. McCusker et al., "Rapid and Extensive Warming Following Cessation of Solar Radiation Management," *Environ. Res. Lett.* 9, (2014): 024005.

266. H. D. Matthews and K. Caldeira, "Transient Climate–Carbon Simulations of Planetary Geoengineering," *Proc. Natl. Acad. Sci.* 104, no. 24 (2007): 9949–54.

267. S. Tilmes et. al., "The Sensitivity of Polar Ozone Depletion to Proposed Geoengineering Schemes," *Science* 320 (May, 2008): 1201–1204.

268. C. H. Trisos et al., "Potentially Dangerous Consequences for Biodiversity of Solar Geoengineering Implementation and Termination," *Nat. Ecol. Evol.* 2 (January, 2018): 475–82.

269. N. E. Vaughan and T. M. Lenton, "A Review of Climate Geoengineering Proposals," *Clim. Change* 109 (March, 2011): 745–90.

270. P. Irvine et al., "Halving Warming with Idealized Solar Geoengineering Moderates Key Climate Hazards," *Nat. Clim. Change* 9 (March, 2019): 295–99.

FINAL THOUGHTS

271. H. D. Matthews and K. Caldeira, "Transient Climate–Carbon Simulations of Planetary Geoengineering," *Proc. Natl. Acad. Sci.* 104 (June, 2007): 9949–54.

272. Michael E. Mann and Tom Toles, *The Madhouse Effect: How Climate Change Denial Is Threatening Our Planet, Destroying Our Politics, and Driving Us Crazy* (Columbia University Press, 2016).

273. World Bank Open Data, "Population, Total - OECD Members, World 1960-2019 [Time series]," dataset accessed April 22, 2021, https://data .worldbank.org.

274. World Bank Open Data, "CO2 Emissions (kt) - OECD Members, World, 1960-2016 [Time series]," dataset accessed April 22, 2021, https://data .worldbank.org.

275. J. Millward-Hopkins et al., "Providing Decent Living with Minimum Energy: A Global Scenario," Glob. Environ. Change 65 (2020): 102168.

276. J. Millward-Hopkins et al., "Providing Decent Living with Minimum Energy: A Global Scenario," Glob. Environ. Change 65, 102168 (2020).

277. J. Millward-Hopkins et al., "Providing Decent Living with Minimum Energy: A Global Scenario," Glob. Environ. Change 65 (2020).

278. "Take Action for the Sustainable Development Goals," United Nations (UN), Sustainable Development Goals, last modified 2022, accessed May 21, 2022, https://www.un.org/sustainabledevelopment/sustainable-development-goals/.

ABOUT THE AUTHOR

Pablo Dias is a professor, scientist, artist, researcher, engineer, and entrepreneur working at the intersection of recycling, environmental engineering, and sustainability.

Pablo's career has taken him from academia to industry, blending research with practical solutions. As a lecturer and researcher at institutions like the University of New South Wales (UNSW), the University of Sydney, Macquarie University, and Universidade Federal do Rio Grande do Sul (UFRGS), he explored environmental engineering, renewable energy, and recycling, teaching students how to assess environmental impacts. His work in life cycle assessment has helped quantify the unseen effects of products, from bottled water to solar panels.

Beyond academia, Pablo co-founded SOLARCYCLE, a technology-driven company pioneering solar panel recycling to create a circular economy for renewable energy.

Pablo is a speaker and advisor on circular economy solutions and climate strategies. His approach bridges scientific rigor with real-world application, making complex environmental issues understandable and actionable. This book was born out of his search for clear, science-based answers to the biggest environmental challenges of our time. It is an invitation to think critically, ask better questions, and take meaningful action toward a sustainable future.

For more information, visit Pablo online at pablodias.net.

www.ingramcontent.com/pod-product-compliance
Lightning Source LLC
Chambersburg PA
CBHW022047020426
42335CB00012B/579